RAND

A Composite Approach to Air Force Planning

Paul K. Davis, Zalmay M. Khalilzad

Prepared for the United States Air Force

Preface

This report reviews a wide range of strategic planning methods and recommends a subset for use by Air Force planners. The research was requested by Major General Robert Linhard, the Air Force's Director of Plans, and was accomplished in the Strategy and Doctrine program of RAND's Project AIR FORCE.

The report also benefited substantially from ongoing work for the advisory board of RAND's National Defense Research Institute (NDRI) , which serves the Office of the Secretary of Defense, the Joint Staff, and the defense agencies.

Project AIR FORCE

Project AIR FORCE, a division of RAND, is the Air Force federally funded research and development center (FFRDC) for studies and analyses. It provides the Air Force with independent analyses of policy alternatives affecting the development, employment, combat readiness, and support of current and future aerospace forces. Research is being performed in three programs: Strategy and Doctrine, Force Modernization and Employment, and Resource Management and System Acquisition.

In 1996, Project AIR FORCE is celebrating 50 years of service to the United States Air Force. Project AIR FORCE began in March 1946 as Project RAND at Douglas Aircraft Company, under contract to the Army Air Forces. Two years later, the project became the foundation of a new, private nonprofit institution to improve public policy through research and analysis for the public welfare and security of the United States—what is known today as RAND.

Contents

Figures

Tables

Summary

After the 1996 presidential election and independent of who is elected, the Department of Defense will probably conduct a major review of national military strategy and the current basis of force planning, the *Bottom-Up Review (BUR)*. Comparable reviews may in the future be conducted every four years, as recommended by the report of the President's Committee on Roles and Missions (CORM). A question for Air Force planners is how best to prepare and contribute. What issues should the Air Force consider, what planning methods should be brought to bear, and when? Also, should the Air Force recommend methods for more general DoD use?

Context for Defining a Planning Approach

Given the preeminent position of the United States, the most important determinant of the future international security environment is the set of choices that it makes about its role in the world, its objectives, and the extent to which to be proactive in shaping the international system. During the Cold War, the United States was more certain about its national security objectives, and it had a grand design. No grand design for the new era has yet jelled. If a consensus emerges on a new design, that design would have a major impact—providing a filter for assessing the relative importance of various kinds of challenges that the national defense strategy must consider. Given this context, as one thinks about approaches to force planning methods, three principles should be kept in mind:

- *Traditional "threat-based planning," the stalwart of defense planning for decades, is no longer an adequate basis for mid- and long-range planning* (see, e.g., Secretary of Defense William J. Perry's 1996 report to Congress). Focusing on a few point scenarios suppresses too many issues.
- *National security planning should instead confront head-on the reality of substantial uncertainty in many dimensions*, political and strategic as well as purely military.
- *Planning activities should be forward looking: They should focus on the long term, and they should encourage examination of options for changing strategy, forces, and doctrine.*

The Principal Activities of Planning

We consider planning to comprise the following seven primary activities, usually thought of as proceeding in sequence, but more typically accomplished iteratively and with a good deal of parallel and intertwined effort:

1. Assessing objectives

2. Recognizing and conceptualizing challenges

3. Conceptualizing and defining alternative response strategies
4. Assessing and comparing the alternatives in terms of effectiveness versus cost and robustness to assumptions
5. Integrating and choosing among alternatives to develop an overall plan
6. Implementing the plan and monitoring events
7. Adapting the plan over time.

Preferred Methods

Drawing on the business literature and past studies, we have identified a set of planning methods suited to Air Force (and DoD) work. There is no single best planning method. Different methods focus on and deal with different generic planning activities. While many of the methods have "planning" in their titles, none stands alone or constitutes a complete methodology. Table S.1 summarizes the methods and relates them to different planning activities.

Methods for Conceptualizing Objectives and Strategies and Recognizing Challenges

Ideally, any overall review should begin by assessing the national objectives and rethinking the National Security Strategy and National Military Strategy. For this, it is important to have some coherent high-level structuring to tie what follows to fundamental national interests and objectives. In Table S.1, reading downward in the second column, the four methods (Uncertainty-Sensitive Planning, Alternative Futures and Technology Forecasting, Out-of-the-Box Gaming, and Assumption-Based Planning) are all powerful for recognizing and conceptualizing challenges, objectives, and broad strategies. All encourage creative thinking to produce *alternatives* rather than a mere rehash of current objectives and strategy. For example, Uncertainty-Sensitive Planning considers not only projections of the future, but also uncertainties associated with *branches* that will be taken at a time that we can more or less predict (e.g., whether the Koreas will or will not unify over the next ten years) and *shocks* (e.g., seriously hostile actions by China in connection with Taiwan) that could plausibly occur at any time. It then calls for a *core strategy*, an *environment-shaping strategy*, and a *hedging strategy*. Or, in a more sophisticated version, it calls for multiple sets of the three types of strategies, with a given set representing a possible explicit or implicit national security strategy (e.g., neoisolationism versus world leadership). Alternative Futures (including technology forecasting) can help significantly in all of this. Out-of-the-Box Gaming, including use of Red Teams, is a time-honored method that can change attitudes about what challenges are real and important. The Day-After games and Revolution-in-Military Affairs games of recent years have been good examples.

Assumption-Based Planning is nominally designed for reviewing plans that already exist. However, it is very useful for *creative* activities, such as unearthing critical assumptions, whether they be of existing plans or organizational mindsets. It can be used early, starting with a critical review of the previous plan. Assumption-based planning encourages environment shaping and hedging actions; it can affect the measures of effectiveness used in analysis; and it defines *signposts* that can be used to monitor developments for indications that the strategy's assumptions are failing.

Methods for Organizing and Managing

Given the broad dimensions of a national strategy, a managerial framework is needed for translating generalities into concrete military missions, operational objectives, and tasks, as indicated in the next several rows of the table. The above methods are useful here also, but for framework setting, we recommend Objective-Based Planning (also called Strategies to Tasks). It produces a hierarchical taxonomy of challenges (objectives, missions, and tasks), after which one seeks strategies and operational concepts for accomplishing them. These should be developed in a joint context and with competition among concepts. Developing such concepts is a creative and cross-cutting activity for which we recommend Concept Action Groups, which bring together "operators" with a joint perspective, Service experts, technologists, and analysts.

Methods for Assessment and Integration

Once alternative concepts have been identified for a given task, there is need for assessment and tradeoffs—systems analysis looking at costs, effectiveness (as measured in a variety of dimensions) versus cost, and various intangibles. Such analysis allows some choices to be made (choices among competing concepts) and provides a base of information for later decision (e.g., curves of diminishing returns for various investments in capability). It motivates the formulation of tentative programs.

The next step is higher-level tradeoff analysis and integration to shape the overall defense program; this should look at the relative priority of missions, operational objectives, and tasks, and do so for a wide range of assumed defense budgets. For this we recommend Adaptive Planning (which includes Capabilities-Based Planning) for relatively comprehensive and sophisticated work involving something we call "scenario-space contingency analysis." This uses an *exploratory*-analysis approach to deal with the numerous dimensions of operational uncertainty. Insights from this can generate Stressful Scenario Sets, which may be used in tradeoff analyses along with curves of diminishing returns to specify "requirements" and manage programs.

A critical element of Adaptive Planning is something we call Strategic Portfolio Analysis, which is needed for balancing investments across such disparate national security objectives as environment shaping, preparation for war-fighting contingencies, and strategic adaptiveness.

Table S.1

A Composite Approach to Force Planning: Products Sought and Methods to Be Used

Product	Methods Useful in Developing Product	Comments
National Security Strategy (NSS) and National Military Strategy (NMS)	Top-Level Structuring	Establishes a coherent framework tied to fundamental interests and top-level objectives.
	Uncertainty-Sensitive Planning (USP) (including looks at alternative strategic environments, budget levels, and views of national interests)	Premium is on open-minded divergent thinking, followed by synthesis. Output of creative and analytic exercises may or may not be clear-cut decisions, but will include insights affecting plan-level decisions.
	Alternative Futures and Technology Forecasts	Focus is on bringing out alternative images of t he future with respect to both the external environment and the national strategy, and with respect to technology.
	Out-of-the Box Gaming	Purposes include thinking the unthinkable and conceiving new strategies.
	Assumption-Based Planning	Encourages creative strategy by critiquing a baseline—by identifying fundamentally important but implicit assumptions that could fail.
Joint Missions and Operational Objectives	All of above techniques	
	Objective-Based Planning (strategies to tasks) conducted for a wide range of circumstances	Premium is on top-down structured analysis. Output is a taxonomy of well-defined functions to be accomplished, motivated by national strategy and its priorities.
Joint Tasks	Objective-Based Planning	Premium is on translating abstract functions into concrete tasks suitable for practical management.
Operational Concepts	Concept Action Groups Comparative systems analysis	Premium is on creative but pragmatic work producing concrete system concepts for accomplishing the various tasks and missions, followed by objective tradeoff analyses to help choose among competitive concepts.
Body of Analysis and Tentative Choices of Approach	Program analysis (especially tradeoff studies and marginal analysis with attention to diminishing-return curves)	Objective is to translate operational concepts into programs for procurement, doctrinal change, training, and so on, and, again, to analyze alternatives.

Table S.1—continued

Product	Methods Useful in Developing Product	Comments
Defense Program and Posture (and, within it, the Air Force program)	Adaptive Planning (which includes capabilities-based planning) using Strategic Portfolio Analysis	Objective is to assess programs and postures, for different budget levels, against a broad range of future contingencies (scenario-space analysis) and against needs to influence the strategic environment and be prepared for strategic adaptation.
	Strategic Adaptation in Complex Adaptive Systems	Objective is to follow a hedged approach initially and to adapt in particular ways in response to specified measures of need.
	Stressful Scenario Sets	Purpose is to simplify expression of require-ments for management of programs and other activities.
	Assumption-Based Planning	Purpose is to review and amend plans to better cope with uncertainty.
	Affordability Analysis	Purposes include providing a life-cycle view of costs, timing major investments to avoid budgetary shocks or temporary losses of capability, and providing reality checks on what can actually be fit into a program with finite resources.
	Organizational-Viability Analysis	Purpose is to assess the competing organizations' ability and willingness to make the changes and investments needed for success of a program (undeveloped).

NOTE: Although the nominal planning process may be linear, the reality is distinctly nonlinear, with many activities in parallel and with many feedback loops. Thus, this table should not be read to imply that a method higher in the table should necessarily be used before a method listed below it. Neither is national military strategy determined in a vacuum. Instead, it reflects a sense of the consequences for operational-level objectives and the defense program. The clear top-down depiction of planning is more realistic in describing results than in describing intellectual processes.

The methods for this are only now being developed. Finally, a key element in defining the overall defense program (and the Air Force program within it) is Affordability Analysis, which helps determine programmatic funding streams over long periods of time. Once a plan is developed, Assumption Based Planning can be used again for review and amendment.

As a final element of assessment and integration, we recommend a hypothesized method called Organizational Viability Analysis, by which we mean an assessment of whether the Services and subordinate organizations involved in a given operational concept are able and truly willing to support the necessary programs and doctrinal developments. This type of consideration has long been important but is usually implicit. Subjective methods should be developed to help formalize judgments.

Although we do not show long-range planning (looking, e.g., 15–30 years out) as a discrete item in Table S.1, it has some special features and needs. Assumption-Based Planning is very useful here. We also recommend a version of Objective-Based Planning (more or less equivalent to what has been called Mission-Pull Planning) that focuses more on *generic* missions, objectives, and tasks—in a variety of operational circumstances—than on scenarios based on current national military strategy and environments. One purpose is to support enabling developments long before their applications can be fully understood.

Finally, let us mention yet another activity not listed in Table S.1, the activity of encouraging organizational change. This requires sensitization, familiarization, changes of basic ideas, and doctrinal work. The Out-of-the-Box gaming methods mentioned earlier can be very helpful. Wide dissemination of the analysis can also be valuable in convincing people at all levels that change is needed (e.g., to cope with actual employment of weapons of mass destruction on the battlefield).

On the Limitation of Methods

To conclude, let us post a caveat: *Strategic planning is an art and craft, not a science*. If undertaken by creative minds, most of the techniques discussed here will do a good job for the Air Force (and the DoD more generally). By contrast, even the most advanced methods will provide little insight if applied in a mediocre way. It is particularly important to apply the methods in contexts, and with "marching rules," that permit and encourage participants to break the shackles of conventional wisdom—not only about current realities, but about what the nature of the future will be, what "good" strategic planners are "supposed" to assume about the future, and what types and levels of forces are allegedly "required." To improve humility on this matter, consider that many long-term planners 20 years ago were subject to conventional wisdom to the effect that long-range penetrating bombers would be obsolete, that carrier battle groups would be highly vulnerable, that ground forces would continue to dominate warfare, and that the Soviet Union would long be a superpower competitor.

Acknowledgments

This study benefited from the ideas of many individuals in RAND, the Air Force, and the business community. We also benefited from a formal review by Professor Paul Bracken of Yale, informal reviews by Maj Mace Carpenter (USAF) and James Dewar, and trenchant suggestions by Glenn Kent. The usual caveats apply.

Abbreviations

BUR	Bottom-Up Review
C^4I/ISR	Command, control, communications, computers, intelligence, surveillance, and reconnaissance
CORM	President's Commission on Roles and Missions
DPG	Defense Planning Guidance
GPS	Global Positioning System
IW	Information warfare
LRC	Lesser regional contingency
MRC	Major regional contingency
OOTW	Operations other than war
PPBS	Planning, Programming, and Budgeting System
RMA	Revolution in Military Affairs
SEAD	Suppression of enemy air defense
WMD	Weapons of mass destruction

1. Introduction

Objectives

Air Force planning—essentially force planning—is a particular example of strategic planning. There is a bewildering range of methods available for use in strategic planning. Each has its proponents; many have distinct advantages; all have limitations. There is no single best planning method, because there are many functions to be performed. *Our principal objective in this report is to describe a composite approach to Air Force force planning in a national context, one that employs appropriate methods for each of the various key functions*. We are familiar with all of the methods presented and have found them for the most part effective for specific planning activities.

Approach

The report begins with simplified depictions of strategic planning at large, drawing upon the business literature and previous RAND work. These distinguish among the many functions that planning (and planners) serve. We also use these models to make certain points that have consistently proven crucial to strategic planning's success and failure in other organizations. Our discussion begins in generic terms, but we then transition into the terminology specifically relevant to the DoD.

After making the transition, we describe a concrete set of methods that comprise a useful tool kit for Air Force planners. We include literature citations, particularly to papers that provide tangible examples of how the methods are used in practice (we give short shrift to methods that seem more notional than real). Next, we discuss criteria for using one or another of the methods. Then we construct a composite approach recommending particular methods for the primary activities we envisage for the next few years of Air Force planning. Some of these should be conducted sequentially, some in parallel. Finally, the appendix describes common pitfalls in planning.

Although the study presents a variety of methods, they all reflect an overarching view:

- The purpose of strategic planning (and, more specifically, Air Force and DoD force planning) is to help the organization prepare for the future by dealing effectively with both profound uncertainties and more predictable considerations.

The approach we propose, then, is suitable for the Air Force seen as a "learning organization" that expects and even looks forward to change as it moves into the future. Much has been written about the challenges of organizational change (e.g., Senge, 1990), but it seems to us that the American military establishment—often depicted as resistant to change—is in fact both willing and able to change markedly, given pressures to do so. Indeed, there is widespread interest in learning and change at all ranks. That does not mean that all the necessary changes will occur (pressures and leadership are critical), but it is a basis for optimism.

Strategic planning is a controversial activity (Mintzberg, 1994). Depending on how it is conducted, it may be distinctly detrimental, harmless but expensive, marginally useful, or exceedingly useful to the organization. In our view, the most useful kinds of strategic planning are creative and integrative, rather than highly structured, rigorously analytic, and deductive. They deal with the creation and adaptation of strategy rather than its routine implementation. This, therefore, is what we emphasize in the current study. To put it differently, we are most concerned with what goes *into* the DoD's Planning, Programming, and Budget System (PPBS), not how the PPBS itself operates.

2. Background: Generic Models of Planning

This section discusses generic models of planning, drawing on the business literature and prior RAND and DoD work. Although some of this material will seem elementary—and even pedantic—to those already involved in planning, it provides a background against which to describe and contrast the various planning methods. The abysmal record of strategic planning in business also suggests that not everything that seems elementary is in fact well understood and assimilated. Nations also have mixed records for strategic planning. The Soviet centralized planning system is now notorious, of course, but we might think also of France's military strategy prior to World War II or American strategic planning prior to the Korean War and in the early stages of the Vietnam War. There have also been several near misses. For example, military planning for the Persian Gulf did not reorient toward defense against Iraq until 1989. Overall, humility is appropriate, and the challenge is considerable. Let us begin by considering a common but naive approach, critique of which will prove useful.

A Simple, Linear Model

There is a natural tendency to view the development and implementation of strategy as straightforward. Figure 1 indicates a particularly simple and linear view in which there is a process of conceiving, defining, analyzing, choosing, and implementing. The output would be "the plan," which the organization would then follow. This model, however, does not ring true. Why? How can one quarrel with such a logical construct?

Before answering this question, let us mention that Figure 1 is not a mere straw man. It characterizes how many individuals and organizations approach planning—with emphasis on analytically rich and precise input data, ambitious forecasts, selection of measures of effectiveness, and optimization of choice. All of this sounds virtuous at first blush, especially to analysts. Arguably, it *is* a good approach so long as one is merely solving well-defined problems and calling that "planning." Indeed, Figure 1 describes a reasonable approach to many operations research problems.

Recognizing Uncertainty

The first problem with the naive model in more complicated cases is that it treats objectives as "given." Establishing objectives (or translating vague ones into something concrete), however, is often half the challenge. The second problem is that the naive model relegates uncertainty to footnotes and does not even mention adaptation. It may recognize the need to do sensitivity analysis when conducting forecasts or comparing strategies, but the assumption is that one can

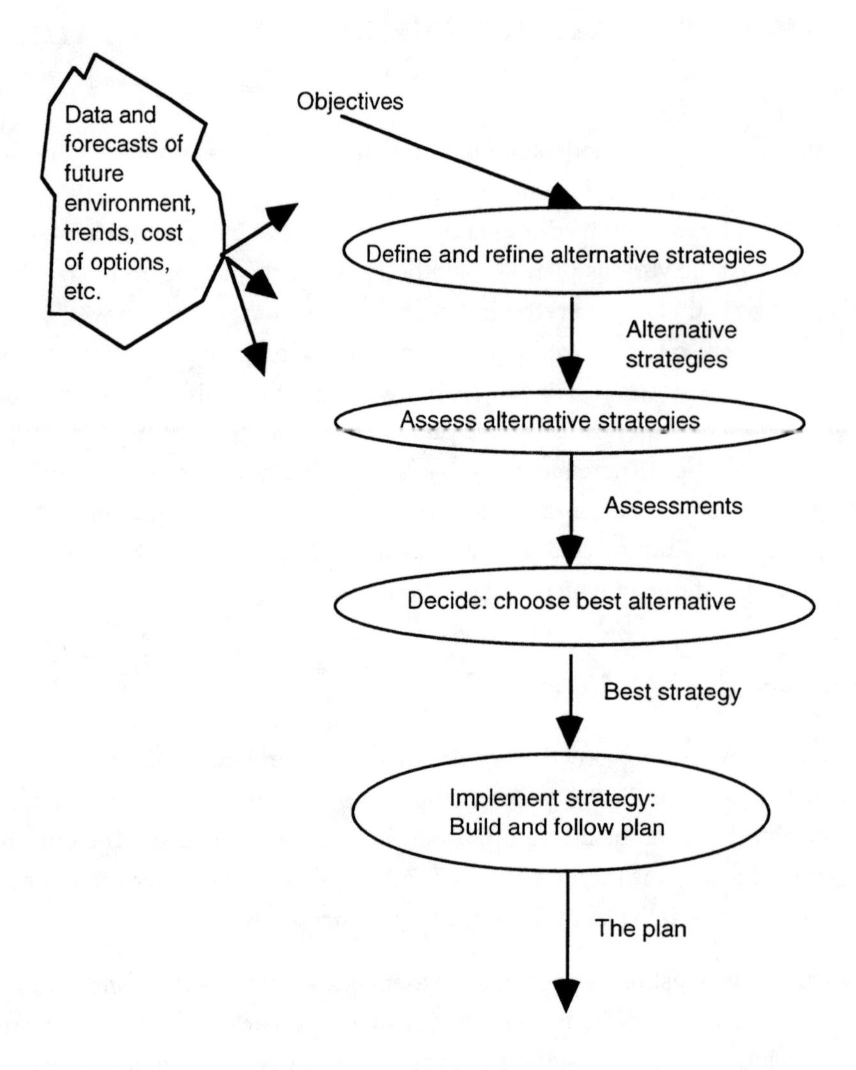

Figure 1—A Naive Linear Model

make reasonable choices and either control the future or count on the environment being stable (in the sense of predictable).

Realistically, uncertainty is fundamental rather than incidental to higher-level planning of organizations in a changing environment (Stacey, 1992; Davis, 1989), including the DoD and Air Force.[1] It is now a commonplace to say that the post–Cold War geostrategic environment is and

[1]There are no good definitions of concepts like "higher-level planning," but we have in mind planning activities associated with what is referred to as establishment of "strategy" and "policy" rather than with, say, the fine tuning or time phasing of programs. More generally, however, in this study we use the term "planning" to embrace activities that extend from policy setting through program development, and even through aspects of implementation. We do not include "comptroller functions," however.

will be highly uncertain. In fact, there were also great uncertainties in the 1960s through the 1980s, but they were often suppressed to simplify planning (a reasonable strategy given a bounding-case threat such as the Soviet Union). Nonetheless, this era is more uncertain in two respects:

- The United States is more uncertain about its objectives and its role in the world than was true during the Cold War.
- Because of the demise of the Soviet Union, the DoD and Air Force can no longer plan forces on the basis of a commonly accepted and massive single threat.

Instead, there are many potential threats to worry about, but no one of them dominates all others or provides a satisfactory basis for planning—even force sizing. *In the current environment, the defense program has to be constructed, explained, and sold on the basis of preparing for a diversity of challenges amidst great uncertainty.*

Many authors have described in the management literature one or another picture of how to categorize uncertainties and the actions that can be taken to deal with them. It is common, for example, to talk about planning narrowing the range of uncertainty—i.e., about *controlling* uncertainty.[2] We should be cautious not to overdo that image, however, because some important uncertainties are not truly controllable. For example, even if the United States had a profoundly insightful long-term strategy toward Asia to shape the environment, and even if it were executed perfectly, China might still end up taking a course we would dislike. China might become a troublesome hegemon, or China, Japan, and Korea might find themselves in a regional competition for military power and influence. The United States can certainly attempt to reduce the likelihood of such dark futures and what we might do to respond to such changes, but we should not imagine that planning for success will assure it. Similarly, we may study the most precise intelligence projections the CIA and DIA can produce, projections that account for economics, demographics, the market availability of modern weapon systems, and so on, but we cannot know with confidence the level of future military threats.

Our view, ultimately, is that, if we contemplate the range of possible situations with some care, we can reasonably hope to rule out some as implausible (e.g., for economic reasons) and to identify some as desirable, some as situations that we could deal with—assuming preparations—and some that would be very troublesome and difficult to deal with. Further, we can reasonably hope to identify actions that would at least alter favorably the *probabilities* of the various situations. Figure 2 sketches the idea. The particular division of the space of possible future

---

[2]It can be argued that many discussions of planning have a virtual fetish about the need to "control" the future and reduce uncertainty (while others ignore uncertainty). There may be profound psychological needs that cause people to ascribe powers to planning that it does not have. Aaron Wildavsky once referred to this: "Planning concerns man's efforts to make the future in his own image. If he loses control of his own destiny, he fears being cast into the abyss. Alone and afraid, man is at the mercy of strange and unpredictable forces, so he takes whatever comfort he can by challenging the fates. He shouts his plans into the storm of life. Even if all he hears is the echo of his own voice, he is no longer alone. To abandon his faith in planning would unleash the terror locked in him." (Quoted in Mintzberg, 1995, p. 203, drawing on Wildavsky, 1973).

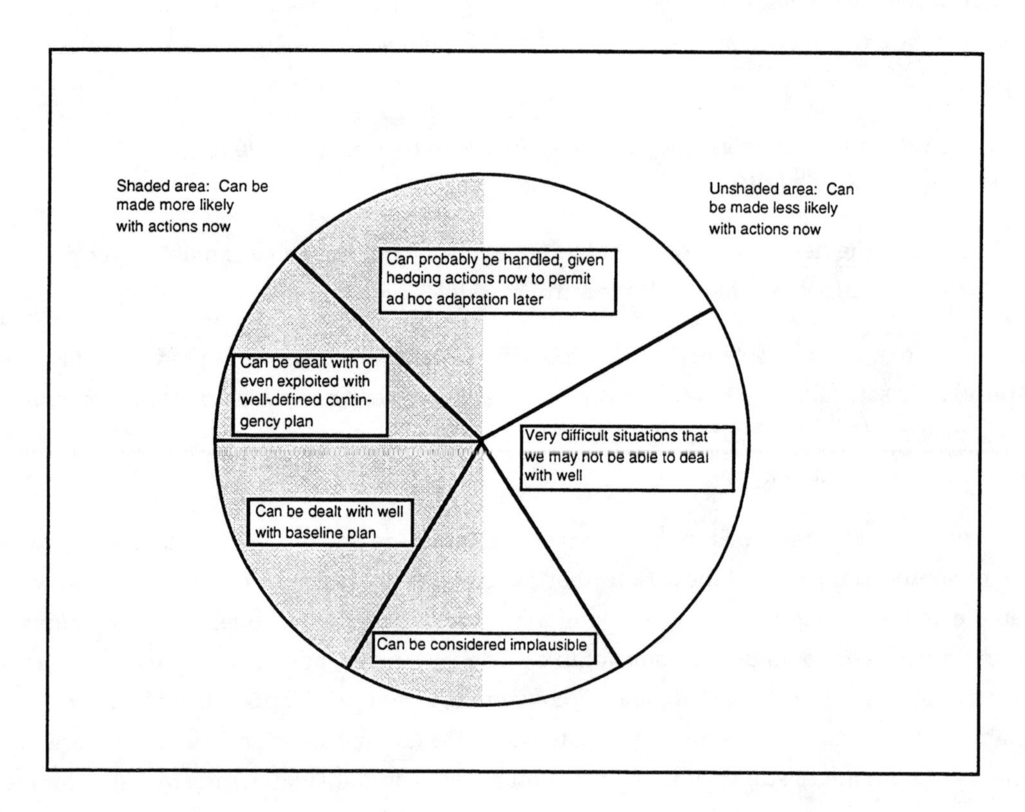

Figure 2—A Notional Breakdown of Possible Future Situations

situations into sectors is arbitrary, but the basic idea is not: The future is not fully controllable, but our ability to deal with change, and even with surprise, can be improved.

With this background, Figure 3 indicates schematically a second model, one that treats objective-setting as a separate step and highlights the need to react to changes over time—not just with "adjustments," but possibly with more drastic changes, albeit changes that might be foreseen by a sufficiently prescient strategic planning system.

In this depiction, strategic planning produces a plan that defines routine adjustments as in Figure 1 (see left branch of Figure 3), but also allows for more discussion of objectives and for defining contingent actions and hedging against surprise (right branch at bottom). Reconsideration of strategy may be necessary.

The distinction between anticipated *branch points* and *shocks* is akin to the distinction between programmed and unprogrammed uncertainties discussed in one of the classic books on organization theory (March and Simon, 1958). It has also been a key element of RAND work to reconsider American grand strategy in the post–Cold War era (Davis, 1989, 1994a; Khalilzad, 1995). The concepts should be intuitively comfortable to most American military officers because

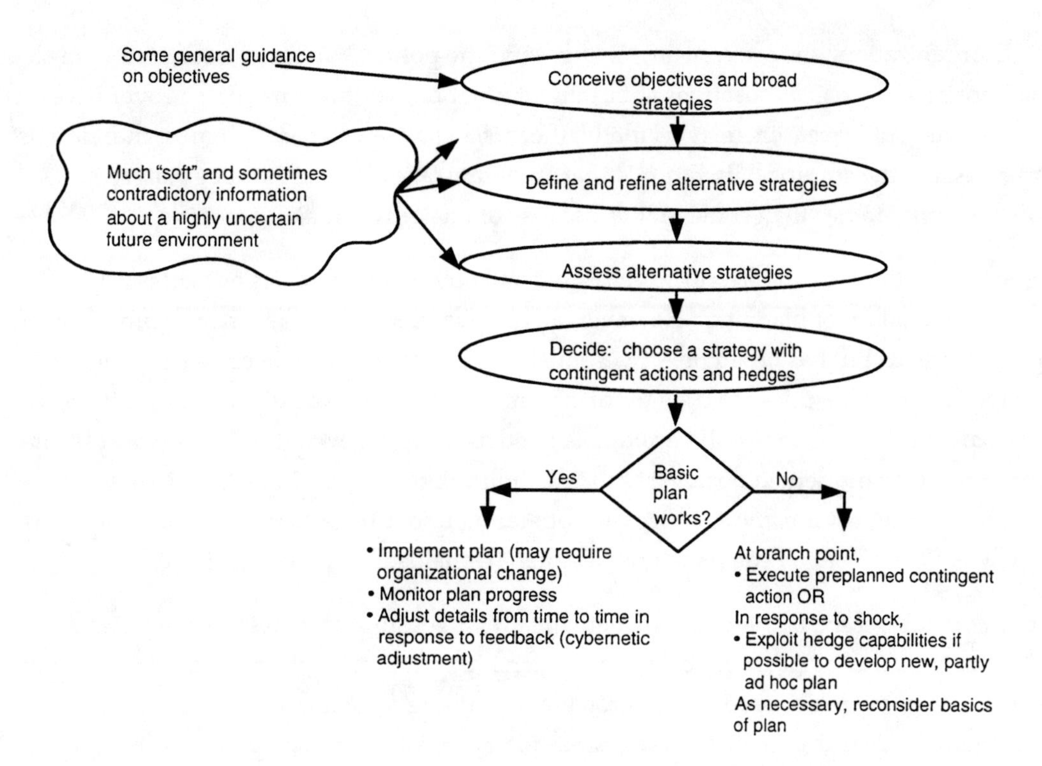

Figure 3—A Model Recognizing Branches and Shocks

there is a tradition (especially at lower levels) of contingent actions and adaptiveness. Culturally, however, the United States is not a nation of chess players; thus, actually planning contingent actions well is only sometimes done, and then more often at the tactical rather than the strategic level.[3]

Learning and Adaptation

Although Figure 3 is more realistic than Figure 1, it still conveys a sense of control or stability by suggesting that most results of uncertainty can be anticipated. The "basic rethinking" branch is, ultimately, a footnote. The model of Figure 3 also retains the notion of a clear-cut decision regarding strategy. At times the leaders of an organization or the presidents of the United States have made and announced clear-cut decisions. However, the organization's leaders may not want to make or announce clear-cut decisions—perhaps because they lack a strong broad vision or because they do not want to make or announce clear-cut decisions for other reasons. Thinking specifically of the president and secretary of defense, these include what some see as the massiveness of uncertainty (i.e., will there or will there not be a global rival or "niche competitor" emerging within the next two decades?), a tactical judgment to avoid taking on frontally those in

[3]Ad hoc adaptation, rather than elaborate and explicit contingency planning, is apparently quite natural psychologically. Experts come to recognize the patterns that call for one or another adaptation. See, e.g., Klein et al. (1993).

the DoD organization who are resisting change, and the political dangers associated with clear-cut decisionmaking (e.g., accusations about threat inflation, hollow armies, or provocative measures that could *create* enemies). Finally, there is a view—well based in most individuals' life experiences and the empirical literature—that an organization's "real" strategy *emerges* opportunistically rather than comes about because of one or two major decisions.

One should not think of opportunistic behaviors and emergent strategies as necessarily representing a failure of planning. However, an organization that lacks a strong broad vision and clear-cut decisions but has broad objectives must proceed by a sequence of heuristically reasonable moves in response to changes of circumstance. So long as the moves are in roughly the right direction, the results will be about as good as possible, even though the process may appear to more plan-oriented observers as mere "muddling along." In fact, some have argued that "muddling along" is *optimal* strategy for operating under uncertainty (Lindbloom, 1968). This view is clearly held by some top businessmen (Donaldson and Lorsch, 1983).

Figure 4 combines some of these ideas schematically, suggesting a process that emphasizes cognitive considerations (learning by senior leadership) rather than clear-cut decisions. In this depiction, one of the major products of strategic planning is a combination of insights, knowledge, and ideas that come to bear later as plans are adapted in response to changing circumstances—often without any claims about choice or change of strategy per se. In this depiction, it may be that outside observers can *infer* a de facto strategy from the nature of the

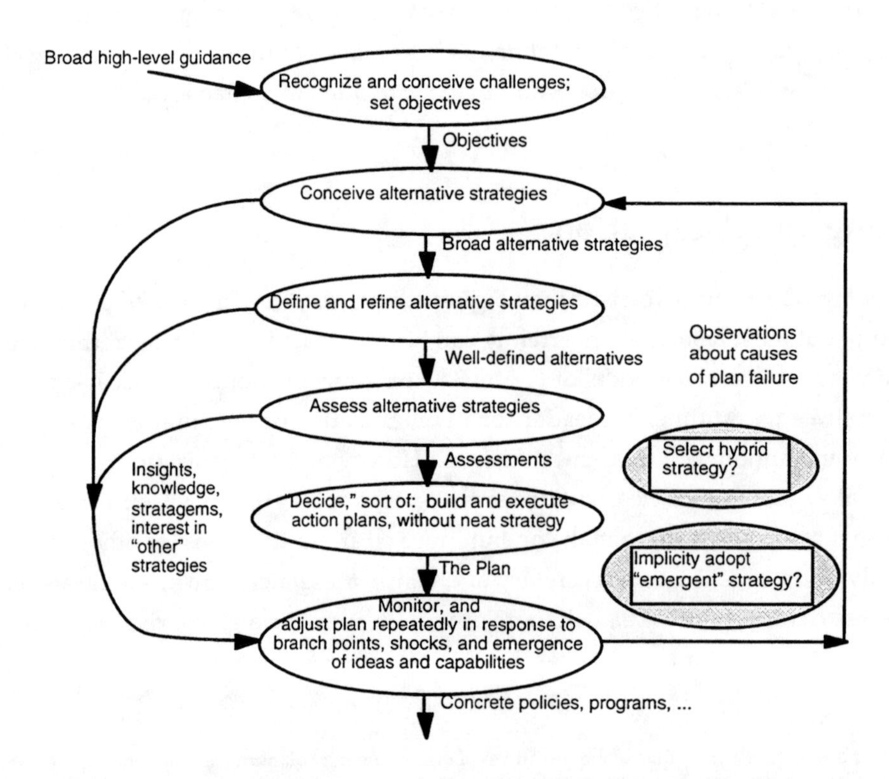

Figure 4—Cognition and Learning-Oriented Model with Muddling Along and Inferred Strategies

actions taken (e.g., the defense program and, within it, the Air Force program), but that strategy may or may not have ever been articulated candidly. In some cases, it may not even have been fully recognized.

Although examples are dangerous because real-world decisions are subject to different interpretations, we shall give a few nonetheless. The first is the unarticulated but conscious strategy of the Air Force in the late 1970s and early 1980s to focus on "rubber on the ramp," procuring modern fighter aircraft in quantity even though that meant temporary shortfalls in readiness. A second and more subtle example of an unarticulated strategy change was, arguably, the increased emphasis given to launch-under-attack for U.S. strategic forces in the early 1980s. This came about through a series of incremental events, such as the decision about how to characterize the survivability of vulnerable Minuteman ICBMs, the decision not to pursue mobile ICBMs, and the decision to increase readiness for nuclear warfighting as part of the deterrent.

Our final example is the decision to proceed with a major Stealth program in the late 1970s. This decision was invisible to most of the official DoD planning system's participants. Further, it was based on judgment rather than rigorous deduction. Different decisionmakers might reasonably have chosen to go in a different direction, one that emphasized ballistic missiles. However, once the decision to proceed had been made, which led to the F-117 and B-2, it *led to* a whole sequence of subsequent decisions, some of them 15 to 20 years later, which together comprise key elements of today's Air Force strategy (e.g., high priority was given to the F-22 fighter, which should provide air superiority for many years). Claiming that this was all foreseen in the late 1970s would be very misleading. The strategy "emerged" in the early 1990s—after the Soviet threat, on which basis the original decision was made, had vanished.

Speculatively, at least, we might think of future emergent strategies involving, e.g., hypersonic aircraft, space-based weapons, airborne lasers capable of boost-phase intercept, or advanced unpiloted aerial vehicles (UAVs).

All of this, then, should contribute to an overall picture of planning with many facets, nonlinearities, and opportunities for the judgment of senior leaders to dominate, either in dramatic and discrete decisions or in a series of incremental decisions over time.

A Comparison Model

As a point of comparison, although we shall not be using it subsequently, Figure 5 shows an integrated model suggested by Mintzberg (1994). This highlights the differences between formulating strategies (an inherently creative activity requiring right-brain skills and a good deal of craft), refining and codifying them (e.g., through the DoD's PPBS), and communicating them (e.g., through the DoD's defense strategy review documents, such as the *Bottom-Up Review (BUR)* (DoD, 1993), including its two-MRC "strategy," as well as through the innumerable program details). Figure 5 also distinguishes among the many different roles played by people called "planners." Planners can provide important inputs to the strategy-formulation process, although

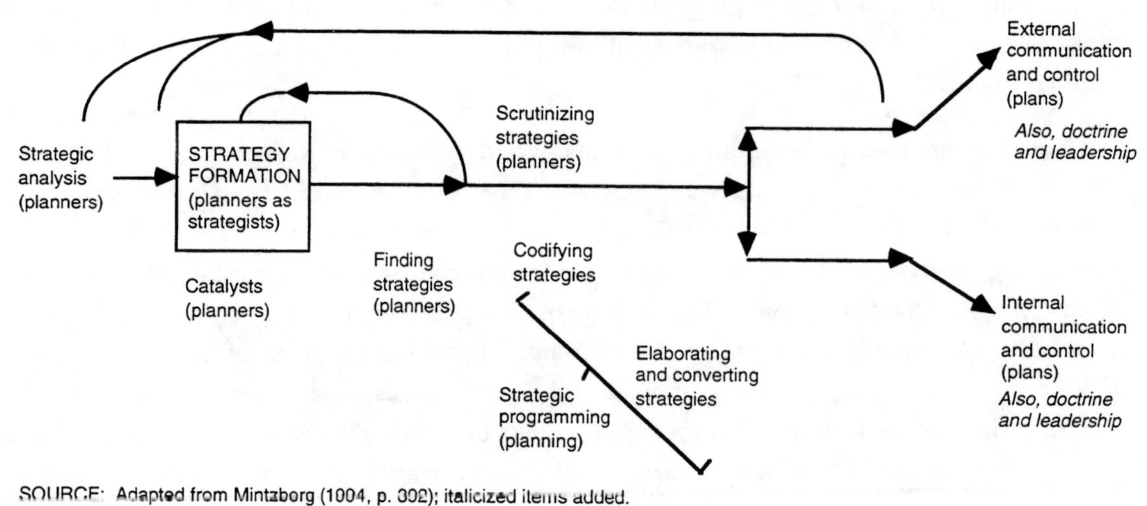

SOURCE: Adapted from Mintzberg (1994, p. 392); italicized items added.

Figure 5—Mintzberg's Overall Model of Strategic Planning Activities

the process itself is fundamentally *not* "analytic." Those few planners who happen to have the requisite talents for conceptualizing and synthesis may also be strategists. Planners play a central role in refining and scrutinizing, analyzing, assessing, and so on. They also play an important role in formulating, monitoring, and adjusting plans over time.

We have added one important element to Mintzberg's depiction, notably the role of doctrine and leadership in communicating and effecting change. Planners can influence this as well. They can inspire organizational change and instruct those willing and able to make those changes. American military leaders actually have many mechanisms available. These include development of new organizational and doctrinal concepts, such as the "Air Force After Next"; high-visibility exercises, such as the Joint Staff's Nimble Vision exercises exploring the value of near-perfect command and control; and conferences. Institution creation is also a means for senior leaders to signal that they are serious about change.

Discussion

Against this background, Figure 6 indicates different phases of planning and some of the many verbs that attach to each. These characterize the differences in planning activities and imply the need for different methods. For example, it is a different matter to conceptualize the challenge and conceive a new strategy than to convince an organization to assimilate it once decisions have been made. Similarly, conceiving new approaches is different from trading off alternative ways to accomplish the same thing within or across alternative strategies. This said, there is a good deal of iteration in the real world, and the best planners are often working on several of the activities in parallel—paying no heed to the rigid sequencing that may be found in their organizations' documents about planning.

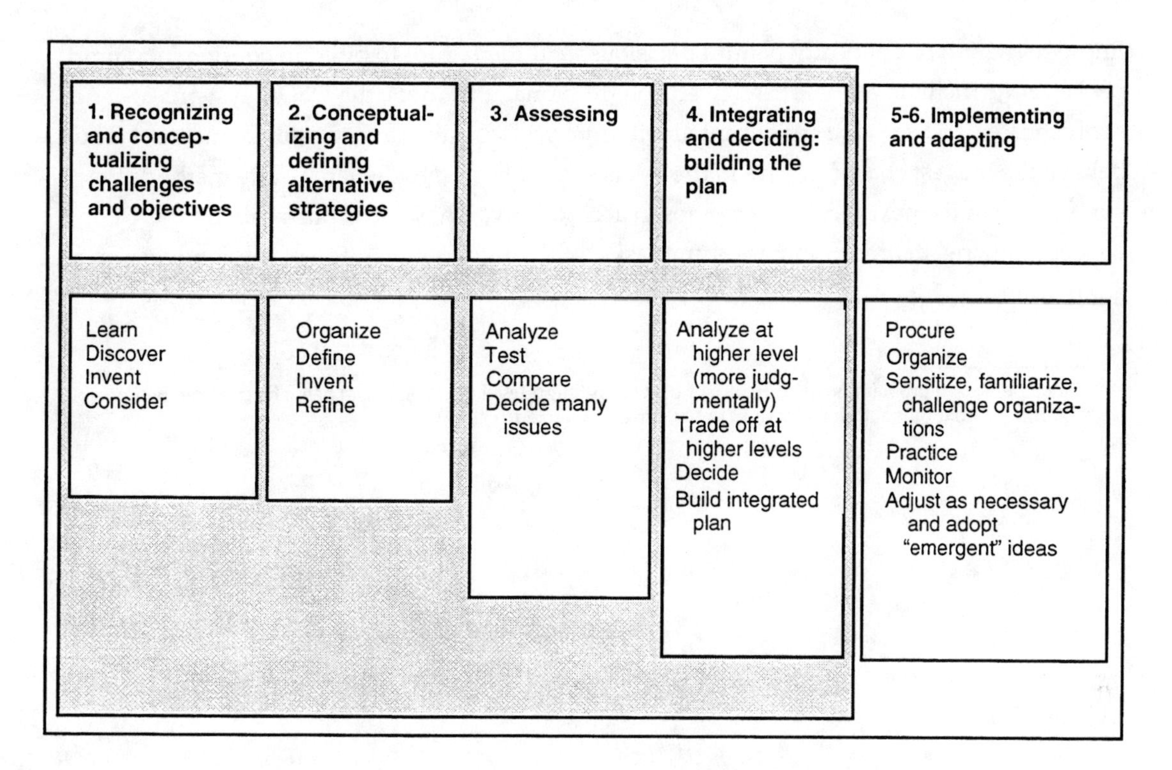

Figure 6—Different Activities for Different Phases

As we contemplate the many aspects of planning, it seems to us that the most difficult planning challenges for the Air Force (and DoD) leadership involve:

- Courageously identifying serious alternatives to the way it currently does business ("getting out of the box")
- Bringing about organizational change (e.g., changes in force structure, doctrine, measures of effectiveness)—without first having to lose a war that would mandate those changes
- Developing future forces and associated doctrine adequate to deal with a wide range of contingencies and contingency details, and, at the same time, effective for purposes of operations other than war (OOTW), environment shaping, and strategic hedging (i.e., hedging against major changes of strategic environment such as emergence of a peer or niche competitor)
- As part of this, making sound cost-effectiveness tradeoffs in a strategic perspective rather than one that implicitly focuses on optimizing around the familiar ways of doing business. In DoD terms, this may correspond to trading force structure away to pay for things in other categories, such as advanced munitions, experimental units, and advanced UAVs.

By contrast, we believe that the Air Force (and DoD) leadership are in much better shape with respect to being able to develop and execute programs given decisions on strategy and meta-issues of program building. Even long-range defense programming and budgeting is in better shape than in earlier decades. Although there are chronic problems of underfunding, optimism, implementation of politically painful decisions, and the failure to build in contingency-fund packages (difficult to accomplish, given congressional attitudes), the DoD's systems are sophisticated and provide significant albeit imperfect visibility to program elements. [4] Thus, we believe the biggest challenges are in the items shown in the shaded area. The methods we summarize in the next section were chosen accordingly. They focus primarily on the conceptualizing, defining, and assessing tasks, with only brief mention of implementation and adaptation.

---

[4]One persistent problem is the difficulty of preventing high-level decisions from being overridden by invisible reprogramming that robs activities thought to have been protected.

3. Force Planning in the Department of Defense

An Overview of National Security Planning Activities

How do the generic pictures of planning described in Section 2 relate to the activities of the DoD and Air Force? Perhaps the first point to be made is that the discussion of Section 2 largely omitted (the exception was discussion of Figure 4) some *informal* functions of strategic planning that are exceedingly important in the national security domain (and also in business settings where strategic planning is effective). Although this report is largely about the formal functions, it is appropriate to step back temporarily and at least note the informal ones. Figure 7 reminds us that there are many different types of national-security-relevant planning besides those of concern to the DoD narrowly. It also highlights the point made earlier, that one consequence of strategic planning is a set of insights, knowledge, and personal relationships that prove important in later decisionmaking, whether in reaction to budget changes or to international crises (see top arrow). This view is consistent with the oft-quoted albeit exaggerated notion that "plans are nothing, but planning is critical." It would be a gross exaggeration to say that plans are nothing when discussing the Air Force or Air Force programs, because they dictate how tens and hundreds of billions of dollars are spent. Nonetheless, there are many important shifts that occur without regard to the prior plan. When the system is working best, those shifts are, however, informed by the insights of past planning.

With this brief and inadequate bow to some very important informal mechanisms, let us now return to the main thrust of the report, which involves the formal aspects of planning.

A Process-Model for Planning Strategies and Resources

General Considerations

Having reminded ourselves of the big picture, let us now adopt a process-model view of the planning processes most relevant here—processes that relate to national military strategy and resource allocation. Figure 8 describes such a model, viewing the Air Force as a major participant and competitor in a process that is fundamentally national and joint.[1] For simplicity, we omit feedback arrows, but it is essential to remember that the real process is by no means linear or purely deductive. The purpose of Figure 8 is to help structure Section 4, which draws on the insights of Section 2 to propose a combination of specific methods for use in Air Force (and DoD)

---

[1]See also Kent (unpublished) and Kent, Crawford, and Bonds (unpublished) for a rich process description that identifies in some detail the types of player and the products to be produced by each activity.

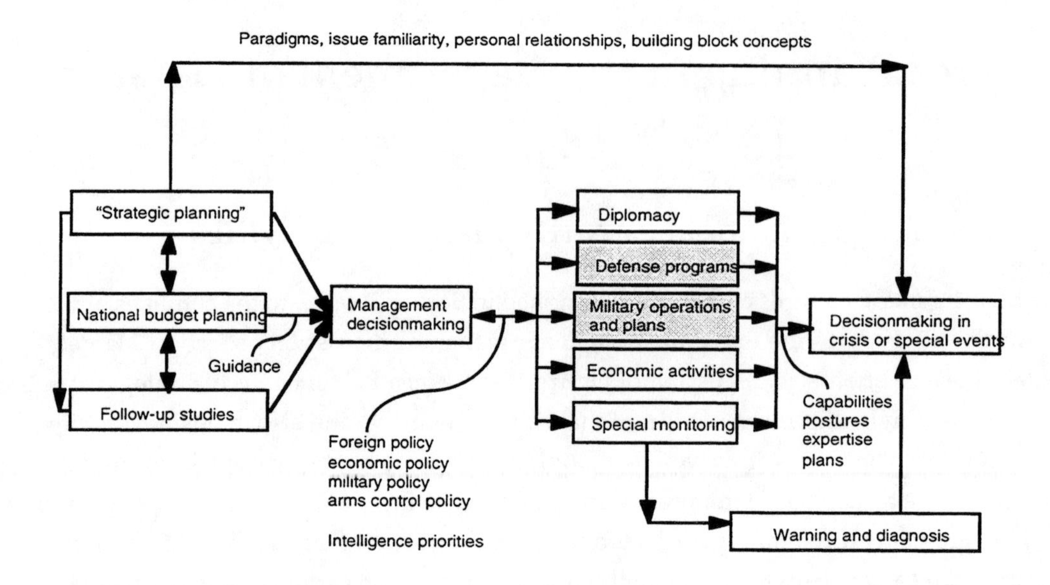

SOURCE: Davis (1989).

Figure 7—An Overview of Planning Activities

planning. Figure 8 summarizes what might be called the "demand." That is, there are demands for methods that help establish grand strategy and help accomplish each of the functions in the figure's bubbles.

Long-Range Planning

Long-range planning (e.g., looking 15 to 30 years into the future, although definitions vary) is a special case to some extent. It proceeds in parallel with midterm planning but cannot be based on a strong concept of the future military strategy or well-defined war fighting scenarios because the uncertainties are too large. Instead, it deals with such issues as laying the base for the future with enabling technologies. We merely touch upon this issue in the current report (for much more discussion, see Smith, 1987, and Murdock, 1994).

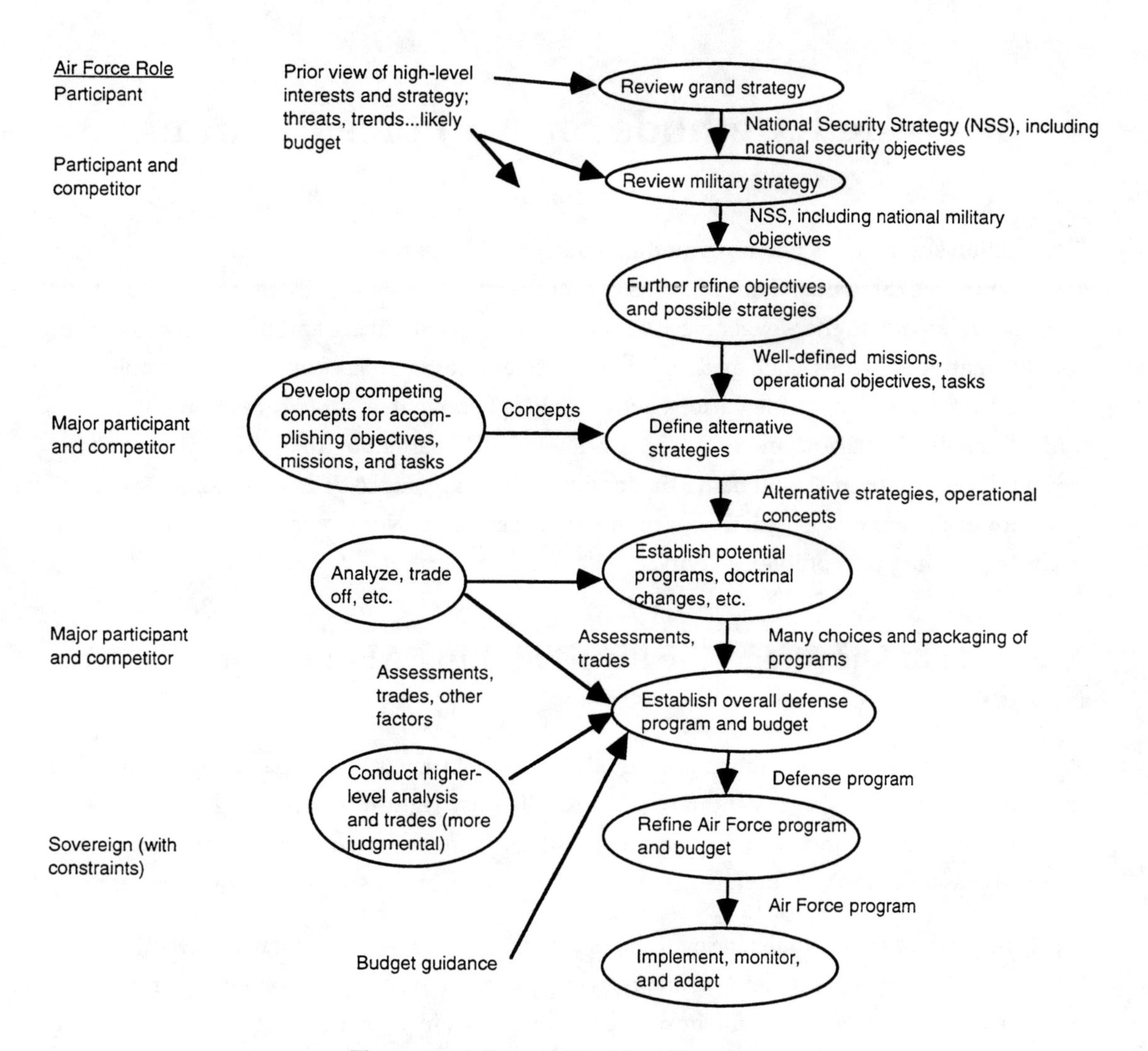

Figure 8—A Process Model of Planning (Iterated over Time, with Various Unindicated Feedbacks)

4. Suggested Methods for Air Force Planning

This section summarizes planning methods that could be of particular benefit to the Air Force leadership. We categorize them in terms of whether they are especially useful for conceptual activities related to higher-level objectives and strategy, structuring and managing (including highly creative activities, such as developing force-employment concepts), or assessment. We then relate the methods to the various activities listed above (e.g., in Figures 6 and 8). Unfortunately, the various methods were independently named by their authors over a long period of time, with the result being that many of them appear to be in competition as "the" way to go about planning. Instead, these are merely techniques. None of them does everything, because planning is a complex activity.

Methods to Encourage Creative Thinking About Objectives and Strategies

A common feature of the methods presented here is that they can help participants with *divergent* thinking—i.e., to help them "get out of the box." This is "right-brain thinking," to use the oversimplified description common in the popular-psychology literature. It is characterized by creativity, association, intuition, leaps, open-mindedness, and a suspension of judgment.[1]

If one purpose of conceptual-phase methods is to bring out ideas, another is to sensitize participants to the existence and seriousness of problems. As a result, some of the methods listed here are also quite valuable in the implementation phase, in which the organization attempts to disseminate ideas, knowledge, sensitivities, and the concept of change being essential.[2]

High-Level Structuring

Although what follows focuses on methods to assist fresh thinking, that work should not be undertaken in a vacuum. There should be familiarity with prior expressions of national interests, objectives, and strategy. There should also be at least some general guidance about the kinds of new issues that might be contemplated. Further, as new ideas emerge about higher-level objectives and strategy, they need to be expressed coherently in a way that ties them to fundamentals, and to past policies. Thus, although "high-level structuring" is seldom thought of as a "method," we see it as an important part of planning. It is, of course, iterative. In a given

[1]Even decisionmaking after exhaustive and rational analysis of alternatives depends on portions of the right brain associated with intuition and emotion. Recent neurophysiological conclusions on this matter are discussed in DeMasio (1994).

[2]For excellent discussions of strategic thinking and operations, see Schwartz (1991) and Peters (1982).

planning cycle, the final results of numerous drafts are ultimately expressed in such documents as the National Security Strategy and the National Military Strategy.

Uncertainty-Sensitive Planning

Late in the Cold War it was becoming evident that major changes in the strategic environment were occurring and that these would necessitate similarly large changes in American foreign and defense policies. RAND developed a methodology for addressing related issues in either a gaming or study-group context (Davis, 1989; Bracken, 1990). Two principal considerations dictated the approach: (1) the desire to confront uncertainty head-on and (2) the desire to encourage a proactive strategy for dealing with uncertainty.

The former was motivated by the need to conceive substantially different strategic environments and strategies; the latter was motivated by the then-current tendency to treat the environment as "exogenous" rather than as something to be influenced. One of us (Davis) had been deeply impressed from lessons learned in cognitive psychology and decisionmaking theory about how successful entrepreneurs are often interested in risk analysis not for the analysis itself (and calculations of alleged success probabilities), but rather as a way to identify the potential obstacles to success, obstacles that should be eliminated in one way or another (e.g., buying out or undercutting a competitor, lobbying for regulatory change, or advertising to create new interests).[3]

The method was used in a study for OSD in 1988–1990; in a 1990 study by both of us for the Commander in Chief, Central Command; and in 1992 when responding to a request from the National Security Council for RAND thoughts about possible courses for U.S. national security strategy. The second application was written up in enough detail to serve as a primer for those who might want to conduct an analogous exercise (Davis, 1994a).[4] We believe readers consulting these references will agree that the results of the exercises were strategic, interesting, and nontrivial.

The key features of the approach are as follows (Figure 9). First, one characterizes the *core environment*, i.e., the "no surprises" picture of the future environment associated with principal trends.[5] Second, one specifically identifies uncertainties of two types: what might be called *branch points* (or scheduled uncertainties, the name suggesting that we know about the uncertainties, can monitor developments, and tend to talk about them in terms of discrete

[3]See the recent book on such matters (March, 1994). See also Donaldson and Lorsch (1983). For an empirical discussion of how real people make operational decisions, drawing heavily upon pattern recognition and intuition, see Klein et al. (1993), a good summary of the "naturalistic decisionmaking" literature, with important implications for decision support and training.

[4]One consequence of this work was that the Bush administration adopted "environment shaping" as an explicit element of its national security strategy. See the "Regional Defense Strategy" of Secretary Dick Cheney published early in 1993. One of us (Khalilzad) was at that time Assistant Under Secretary for Policy Planning.

[5]The "no-surprises future" is by no means the best-estimate future, since the likelihood of no surprises is very small. We thank Jim Dewar for reminding us to stress this point.

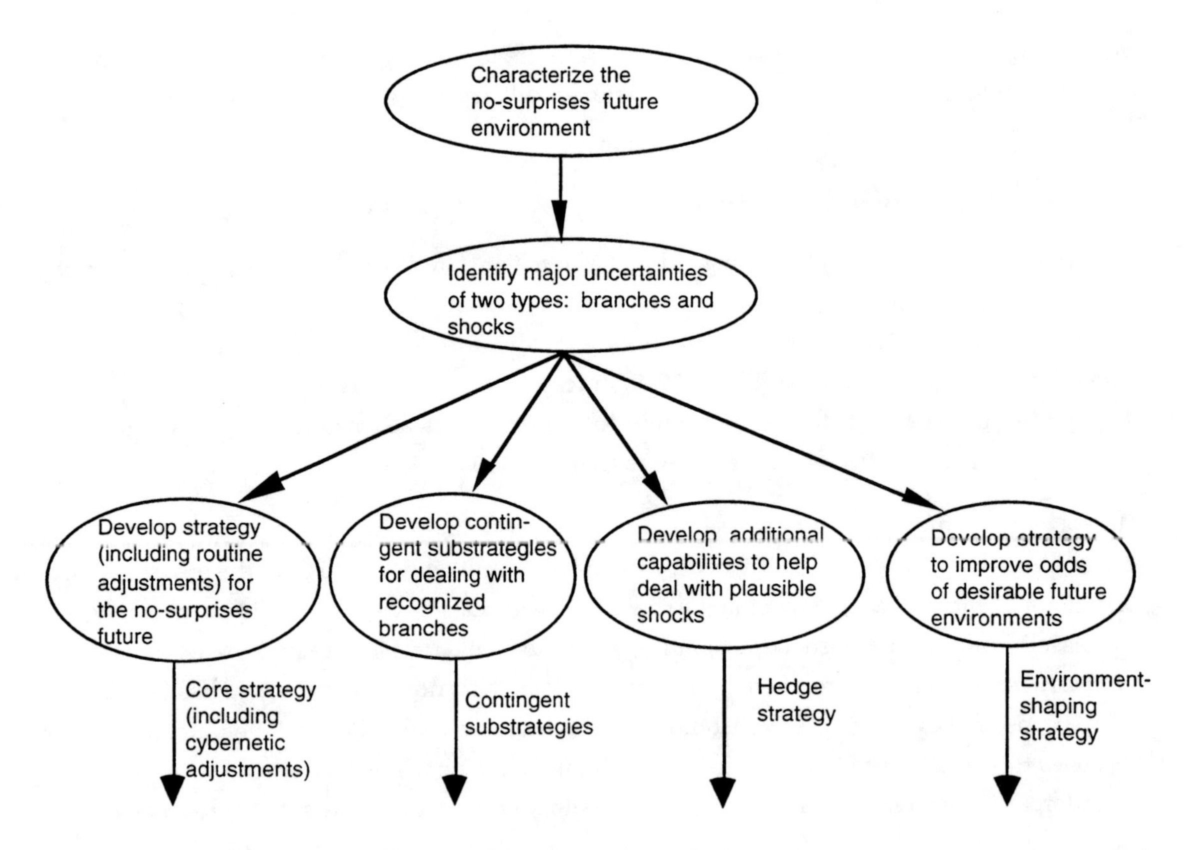

Figure 9—Uncertainty-Sensitive Planning

branches that will eventually be taken) and *shocks*. We knew in the late 1980s, for example, that the future of Germany was a major issue that might be resolved one way or the other within a short period (a branch point), and we could contemplate contingent actions either way. By contrast, there was a long list of potential shocks, both good and bad, that might occur at any time, or never.

Identifying potential shocks is especially critical because getting out of the box requires facing up to the fact that *some* of the long list of allegedly improbable major events will actually occur. Perhaps China will never invade Taiwan and perhaps Russia will never invade Lithuania, but if we consider the long list of strategically plausible events and realize that some of them *will* occur, then our approach to strategy will be different. The 1989 study explicitly listed as important potential shocks the possible collapse of the Communist party and the reunification of Germany. In our March 1990 study, we highlighted the possibility of a threatened Iraqi invasion of Kuwait that would be ambiguous (was it a mere threat or would the invasion occur?).

Once participants are enthusiastically engaged in recognizing uncertainties, they can turn to the development of strategy. One approach here is to have the same group develop three different strategies: a *core strategy*, an *environment-shaping strategy*, and a *hedge strategy*. This can be done interactively with a leader or facilitator at the chalkboard or computer.[6] The real-time output is a set of bullets and comments, but this can later be turned into something more organized and respectable. The intention, of course, is that the core strategy represent the key elements of a no-surprises extrapolative strategy, whereas the environment-shaping strategy should include actions to influence the behavior of other nations (or, for example, the development of new technologies). The hedge strategy consists of actions taken and capabilities developed to prepare for ad hoc adaptation at the emergence of shocks.

In a variant of the approach (e.g., Davis, 1994a), teams may develop alternative grand strategies (each with core strategy, environment-shaping, and hedging components) representing different mindsets (e.g., America-first and nonintervention on the one hand versus engagement and protect-the-Great Transition on the other). Follow-on analysis can array the strategies against each other, comparing and contrasting as appropriate. In some cases, this can produce a hybrid strategy with many of the best features of several "pure" approaches. We recommend against treating "pure" strategies as real alternatives for decisionmakers. Typically, they are merely clarifying devices, not real options, because in a democratic government "real" strategy should usually accommodate a variety of views (see also Donaldson and Lorsch, 1983).

Alternative Futures (a.k.a. Future-Oriented Scenario-Based Planning) and Technology Forecasts

Current business planning often includes a version of scenario-based planning in the form of exercises in which participants define alternative futures (e.g., with different demands for certain types of products, different regulatory environments, and different levels of competition). The challenge is then to develop a business strategy that would be robust enough to cope with all of the most plausible futures. This approach has deep roots, which go back to the nuclear strategy efforts of Herman Kahn in the 1950s and, in later years, the Hudson Institute. There are some good published descriptions of the approach, although with few details (see especially Thomas, 1994). A classic is the work by Shell Oil (Wack, 1985), which describes how Shell's strategic planning helped anticipate the oil shock of 1973 in general terms, allowing Shell to move decisively when it occurred (and, later, after a predictable oil glut developed). Within the defense realm, there have also been many examples of alternative-futures work. One RAND example is Kugler (1995), which describes alternative strategic futures ranging from a relatively benign world to one with a virtual explosion of conflicts, many of them based in ethnic or religious disputes. Another is Khalilzad (1995), which discusses alternative U.S. grand strategies—looking not only at alternative external environments, but also at alternative national security strategies, such as neoisolationism, sharing leadership with others, or maintaining U.S. global leadership and precluding the rise of a global rival.[7] Yet another approach involves building alternative

---

[6] Methods for group brainstorming and decisionmaking are included in Kleindorfer et al. (1993), a good survey of the decision-sciences literature, including work from the cognitive sciences.

[7] An alternative-futures approach was used in Spacecast 2020 and is being used in the current Air Force 2025 study concerned with the Air Force After Next.

cognitive decision models to help understand how the leaders of other countries might be thinking, individually or collectively.[8]

Figure 10 is a schematic of the method. It makes the point, seldom mentioned, that the alternative futures considered should vary both the external environment (e.g., will China arise as a hegemonic threat in East Asia?) and the internal environment (e.g., will the United States fall into a kind of neoisolationism?). Depending on the combination assumed, different sets of "requirements" would be seen, and different sets of capabilities would be sought. In any case, good strategy might consist of establishing programs that would be relatively robust across the alternative futures. That, however, is an ideal. If the futures are sufficiently different, strategy will have to tilt significantly: One cannot prepare for all futures. In that case, good strategy might nonetheless lay the basis for subsequent strategic adaptation in other directions.

Technology forecasting is an essential ingredient of strategic planning because, without it, planners will often not appreciate the potentialities of what is emerging. An example of this can be found in the "New Vistas" study by the Air Force Science Advisory Board (USAF, 1996). Such work (and analogous work by, e.g., the Defense Science Board) may address such diverse issues as the significance of distributing processing, information warfare, technology for improving strategic mobility, and directed weapons in space. The challenge is for the participants to relate technological developments to possible military functions. In that respect, some of the best participants are also those who should be involved in concept action groups as discussed below.

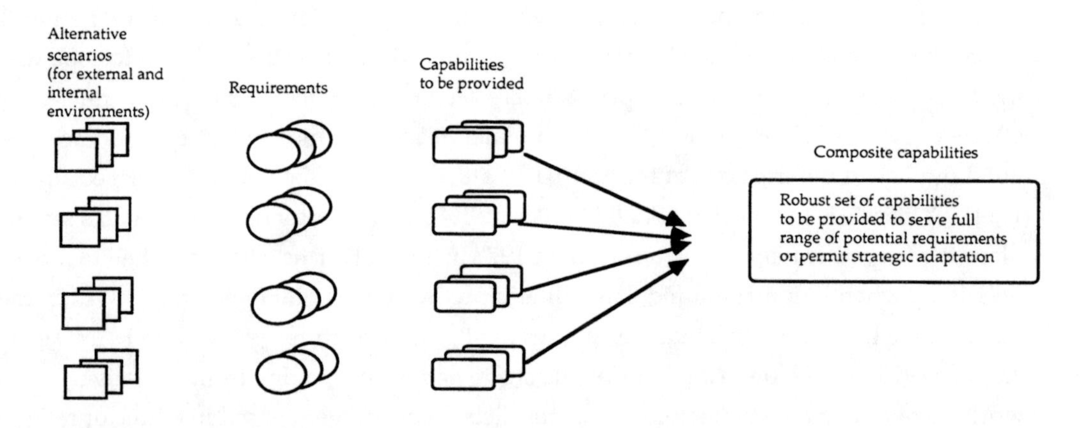

Figure 10—Building Capabilities to Deal with a Variety of Potential Futures

---

[8]Cognitive decision modeling has been applied primarily to crisis issues (Davis and Arquilla, 1991), but there have been some applications to more strategic issues involving counterproliferation (Arquilla and Davis, 1994) and negotiating with North Korea (Davis and Khalilzad, unpublished).

Out-of-the-Box Gaming

A broad class of valuable methods is what we call "out-of-the-box" gaming—gaming to look freshly and creatively at challenges.

The Day-After Games. A particularly effective method for "thinking the unthinkable" with respect to crises and wars has been the "Day-After Game" introduced by RAND's Roger Molander, Dean Millot, and Peter Wilson in work for the Air Force. Conceived to sensitize participants to the possibility of the United States actually finding itself in a nuclear crisis, and to the extraordinary dilemmas that would arise, the authors constructed an efficient half-day game with several moves. The same approach has subsequently been used to explore more general use of weapons of mass destruction (WMD) and, most recently, information warfare. There have been dozens of these games held at RAND, in the Pentagon, in Europe and Russia, and elsewhere. A recent OSD-sponsored information-warfare game included very high-level participants from throughout the national security community and other agencies, such as the Department of Commerce and the F.B.I.

By and large, the usual format is to (1) break into teams; (2) present the teams with a crisis situation that includes the possibility of WMD use and allow them to debate the issues for about an hour; (3) present the teams with a new situation somewhat later in the crisis, after WMD have in fact been used (i.e., the players are no longer able to remain in the secure and abstract language of deterrence), again allowing about an hour's discussion and decision; and (4) bring the teams back to "today," asking them, with the benefit of their new appreciation of nuclear crisis, to identify policies and strategies that might be undertaken today to avoid some of the dilemmas of tomorrow they had just experienced.

The Day-After games have been credited as a substantial factor in causing numerous uniformed officers and civilians throughout the national security community to take seriously the necessity of planning actually to operate in a WMD or information-warfare environment.

The basic methods have been documented as part of the initial nuclear-focused series (Molander, Millot, and Wilson, 1993). Although they appear simple, their success has been highly dependent on the sophistication and preparation of the game designers, the quality of the participants, and follow-on activity to make sense of results and draw generic conclusions.

Future-of-Warfare Games. To provide a getting-out-of-the-box method that focuses more on military issues, we suggest future-of-warfare games akin to those conducted over the last several years by colleagues Bruce Bennett, Daniel Fox, and Sam Gardiner as part of the continuing effort by OSD's Office of Net Assessment to explore the potential implications of what some (including ourselves) see as the emerging revolution in military affairs (RMA).[9] A good short account of this

---

[9]There have been a number of interesting efforts in the RMA work, including gaming and other exercises led by David Andre, a consultant for SAIC, and Mark Herman of Booz-Allen & Hamilton, Inc. For a good summary of conclusions, see results of the Office of Net Assessment's summer study at the Naval War College (1995), including a paper by Michael Vickers. There is also an ever-increasing literature on the RMA, which we shall not cite in detail here.

work is Bennett, Gardiner, and Fox (1994). To convey a sense of the method, let us describe the first (in 1991) of what turned out to be a lengthy series of games. Red Team participants gathered in a war-gaming room replete with maps and other paraphernalia of gaming, such as high-level order-of-battle data. They were told that they were the general staff to Saddam Hussein, who was seeking a new operational strategy for invading Kuwait and Saudi Arabia. They were then shown the simulated results of the plan drawn up by their predecessor general-staff members (who had been shot). Basically, that plan was a marginally upgraded version of the one used in 1990, extended to include a dash to Saudi Arabia. Although the invasion caused problems for the United States, the eventual result was complete destruction of the Iraqi army, due primarily to the awesome lethality of American air forces. The new general staff, obviously, was expected to do better.

Players used a variety of methods to identify U.S. vulnerabilities and Iraqi vulnerabilities, and to track backward from those to possible tactics. They were encouraged to "think big," which meant being able to broaden the arena of conflict and to include purely political actions. Some of the tactics generated would indeed cause the United States great difficulty in attempting to respond. However, in a subsequent version of the game, U.S. players began to identify counters to those tactics. And so on through a series considered to extend over time well into the next century.

Some of the conclusions are now relatively familiar. Highly accurate long-range fires (whether from aircraft or missiles) may be dominating the nature of modern warfare. Even second-rate nations may have relatively accurate weapons and good intelligence, in part by exploiting peacetime commercial space systems. That in turn will force antagonists to disperse forces, concentrating only in quick bursts of movement. Antagonists may also seek shelter in urban sprawl (or jungles and mountains). There may be an increased premium on the use of infantry, and on covert deployments (e.g., from commercial roll on-roll off ships or trucks). There may be extensive use of WMD, particularly chemicals, and threatened use of biological and nuclear weapons, primarily for deterrence (e.g., to deter a U.S. counteroffensive that would seek to destroy the aggressor's army and occupy its territory). And so on. Our point here is that the games have been effective in encouraging participants to envision new concepts of operations, weapon systems, and higher-level strategies—including strategies involving strategically important "events" within the United States, information warfare, and so on.

Red Teams and Asymmetric Strategies. Many other approaches fall under the general rubric of out-of-the-box gaming and analysis. We have already mentioned Red Teaming, the essence of which is to create a team of people charged with defeating our forces and strategies, perhaps by adopting strategies toward which our adversaries would be more inclined for strategic, cultural, economic, and military reasons. Recently, interesting work of this type has gone under the rubric of developing *asymmetric strategies*, which often involve relatively low-technology methods that nonetheless could cause grave difficulties for our forces. See, for example, Watman et al. (unpublished) and the 1995 summer study of the Defense Science Board. Currently (mid-1996), RAND is using operational-level gaming in a major analytic study for the Air Force directed by

Natalie Crawford. As so often has happened in the past, the insights gained are remarkable, due especially to use of red teams and the search for clever, asymmetric, strategies.

PATH Games. The last gaming method we shall mention here is a form of gaming in which teams build defense programs and assess their adequacy as a function of time. That is, the games examine the *path* from now until then. Such games were useful during the 1980s when thinking about how the United States and Soviet Union might transition their strategic forces from offensive dominance to defensive dominance. PATH games were invented at RAND in the 1950s, and there is some documentation of the original procedures, as well as those used in so-called SAFE games. For a more recent report, see Finn (1988).

Assumption-Based Planning

Assumption-based planning has many of the same features as uncertainty-sensitive planning and was developed more or less contemporaneously (although its origins trace back to work by Builder in the early 1980s). It is tailored, however, for use in *reviewing* plans rather than for use in the initial development of alternative strategies. It has proven valuable in reviewing and rethinking both explicit and implicit organizational plans. The basic concept is for a group to think as deeply and creatively as possible about where the fatal flaws might be. What might conceivably happen that would obviate the plan? What deeply buried assumptions are so implicit as to be usually invisible, but are in fact fundamental and uncertain? The answers may range from the geostrategic (e.g., that the United States will not have major border-protection problems) to the operationally concrete (e.g., that the United States will not have to conduct counteroffensives through urban sprawl). From such answers can be synthesized a set of recommended changes to watch for failures of assumption (identification of *signposts*), to hedge against such failures, and to influence the likelihood of failures (Figure 11).

The approach is well described and illustrated (Dewar, Builder, et al., 1993), drawing upon early-1990s work for the leadership of the U.S. Army (Dewar and Levin, 1992). The work involved review of the AirLand Battle–Future (ALV-F) concept. A top Army leader praised the work and its influence in extraordinarily strong terms in response to a RAND survey obtaining feedback from Army sponsors. It has subsequently been applied within RAND for a variety of projects and is now being used to help the Army evaluate Force XXI.

A key step in the assumption-based planning process is identifying which assumptions might break down and, if they did, whether enough warning time would be available to take actions at the time. Figure 12 sketches the logic tree here. Clearly, if the failure of the given assumption might occur with too little warning time for action at the time, then actions should be considered in the near term. This is straightforward reasoning, but it is not always so easy to apply. In the context of Air Force planning, an illustrative issue involves assumptions about whether aircraft with little or no stealth will become vulnerable over time. How quickly might F-15Es become vulnerable to regional-threat air-to-air or surface-to-air missiles? Would there be time to adapt, or is this concern a decisive argument for the F-22? In what numbers?

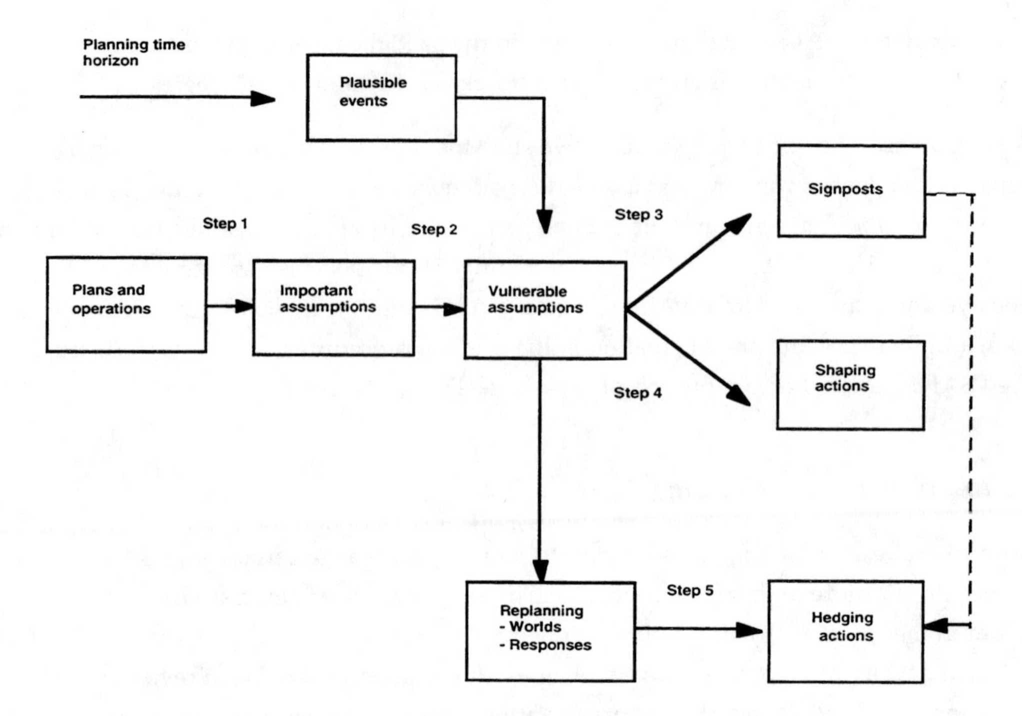

Figure 11—A Schematic of Assumption-Based Planning

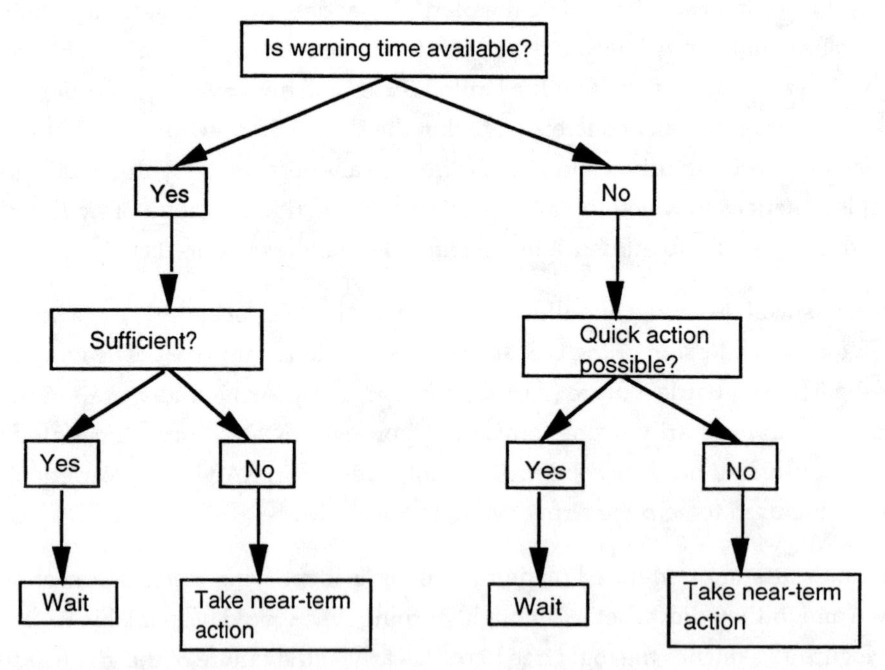

Figure 12—Decision Tree for Thinking About Whether to Take Hedging Actions Now

The authors of the assumption-based technique compare their methods with others and argue that it is particularly useful in dealing with a future containing fundamental uncertainties about an organization's ends. Our own view is that assumption-based planning can be quite useful not just in reviewing plans, but throughout the conceptual stage of work (where the previous plan can always be taken as a baseline against which to apply the method) and in helping to establish measures of effectiveness.

Methods for Structuring and Managing

Assuming reasonable statements of high-level objectives and strategy, there is still a great deal to be done before one can have anything like a real "plan." Indeed much of the work at the level of policy and strategy is quite abstract. One of the most important challenges for planners is translating the abstractions into concrete jobs to be done and then finding ways to do them well and economically. Further, there is need for an overall hierarchical architecture relating the various jobs to be done. This is not just an intellectual effort; it is also essential for structuring and managing the resulting programs and other activities. Objective-based planning is a powerful method for this purpose.

Objective-Based Planning

In 1988, RAND produced the objective-based planning methodology (usually called strategies to tasks) in a paper by Ted Warner and Glenn Kent. The methodology has evolved considerably since then (e.g., Thaler, 1993). A key feature of the method was its emphasis on objectives and being top-down. As described in Kent and Simons (1994),

> The concept centers on a subordination of objectives whereby outlining a plan for attaining stated goals at one level of organization defines objectives to be achieved at subordinate levels of implementation. It describes a process by which one may proceed coherently from stated national security objectives, to national military objectives, to regional campaign objectives, to operational objectives, and finally to military tasks. The process provides a clear audit trail from top to bottom, gives clear meaning to plans of action (strategies) formulated at each level, and offers a certain stability for our national security planning, year by year and era by era. The concept sets the stage for a process of allocating national defense resources to best effect and could be applied to the DoD's Planning, Programming and Budgeting System (PPBS).

Figure 13 (Pirnie and Gardiner, 1996) describes objectives-based planning in a way that emphasizes its top-down perspective. Figure 14 shows the first levels of detail (Bonds, Hura, et al., 1995). In practice, objectives-based planning is not purely top-down. Instead, the framework that emerges is usually the result of numerous iterations that reflect familiarity with important low-level tasks and a variety of operational circumstances. As with many top-down planning constructs, that which is "top-down" is often to a large extent a means for pulling things together coherently after one concludes what is and is not to be stressed.

Examples of the kind of generic operational objectives that emerge from this work are (1) establish an effective coalition, (2) achieve superior operational awareness, (3) force entry into a region, (4) enhance capabilities of U.S. allies, and (5) sustain forces. Tasks are concrete as in

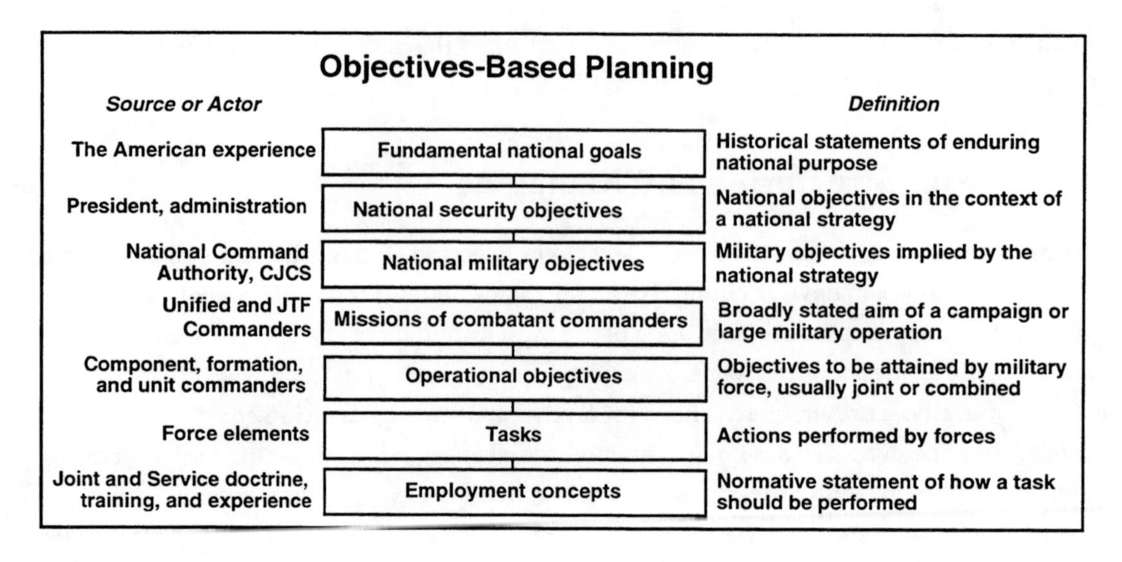

Figure 13—Objectives-Based Planning

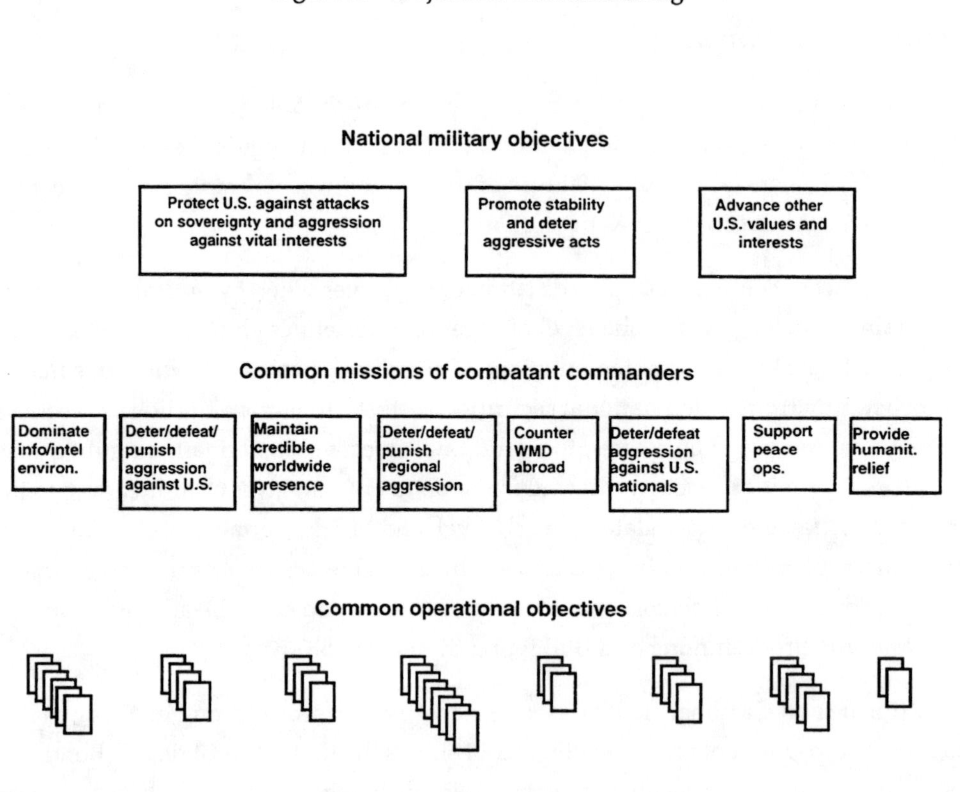

Figure 14—Joint Missions and Operational Objectives

"repeatedly close runways to suppress sortie generation." The approach is relatively sterile unless one actually gets down to the level of tasks (see below for examples).

If application of the approach leads to the determination of military tasks, these in turn point to the concepts, weapons, and support systems that may be needed. This methodology encourages the discussion of alternative means for accomplishing missions, operational objectives, and tasks. Different services may compete for performing the same task or accomplishing the same objective.[10]

Mission-Pull Planning

By and large, the usual version of objectives-based planning makes the most sense for mid-term planning, because—as usually described—it makes use of the current national military strategy and the planning scenarios that it provides. However, it is difficult to place much credence in forecasts of interests, objectives, missions, and the like, much less particular scenarios, when talking about 20 to 30 years in the future. Instead, it makes more sense to focus on generic operating environments and generic missions and tasks, with less emphasis on establishing a clear audit trail from current national military strategy to tasks and programs. Clark Murdock in Air Force Long Range Planning (previously in OSD's Policy Planning office) and Wade Hinkle, now at the Institute for Defense Analyses and previously in OSD's Policy Planning office, have proposed "mission-pull planning" to pursue this line of attack. Figures 15 through 17 suggest some of the features notionally. Figures 15 and 16 show how they propose to decompose the problem into tasks and subtasks; Figure 17 illustrates a scorecard method for characterizing capability.[11]

There is a great deal of overlap between mission-pull planning and objective-based planning: both involve taxonomic decomposition of higher to lower objectives, the lowest being tasks and subtasks; both use scorecards for some aspects of evaluation; and so on. In essence, except for differences of terminology and the like, mission-pull planning is what objective-based planning becomes if one focuses on more generic operational missions and tasks.[12]

This said, one important contribution of the mission-pull work has been emphasizing the diversity of possible situations (operating environments). We recommend that those using objective-based planning recognize *explicitly* that the various operational objectives and tasks must be accomplished in a variety of circumstances varying with respect to political-military context, terrain, weather, and so on (see also Davis, 1994b and the discussion below of adaptive planning). If this is done, then there is considerable convergence.

---

[10]For current descriptions of the methodology, readers should refer to unpublished work by David Ochmanek and Stephen Hosmer; they may also wish to contact colleague Glenn Kent directly in the RAND Washington Office. Some of the current work has been directed by Leslie Lewis, John Schrader, and others for the Joint Staff's J-8; some has been conducted for the Air Force by Tim Bonds, Myron Hura, and others. Bruce Pirnie and Sam Gardiner have applied the approach to OSD work. The best single point of contact, however, is Kent himself. See also a recent application to information warfare by Charles Heimach, working for ANSER Corporation (available through the Office of Net Assessment).

[11]See Murdock (1994) for discussion. There have been no in-depth documented applications to date, but the ideas have been widely briefed (Murdock, 1995; Hinkle, 1994). We believe their essence can be incorporated readily in a variant of objective-based planning.

[12]One of us, at least (Davis), believes that the top-down aspects of objective-based planning (Figure 13) have been overemphasized at times, obscuring the fact that the "real work" of the approach has typically been at the lower levels—identifying the operational objectives and the many specific tasks that must be accomplished. In this view, the differences between objective-based planning and mission-pull planning are less than others might think.

Operating Environments (notional)	**Mission Areas and Missions**
Humanitarian aid Noncombatant evacuation Counterterrorism Coup de main Rural-based insurgency Urban-guerilla warfare Peace enforcement Peacekeeping Counter-WMD Regional defense, WMD environment Blue-water MRC/LRC Littoral warfare MRC in a WMD environment Homeland defense against full threat spectrum	Defend the United States and its citizens • Deter or defeat WMD attacks on United States • Combat terrorism • Evacuate Americans from overseas ... Deter, prevent, or defeat proliferation or use of WMD or the means of delivery ... Defend regional allies, friends... ...

Figure 15—Operating Environments and Missions in Mission-Pull Planning

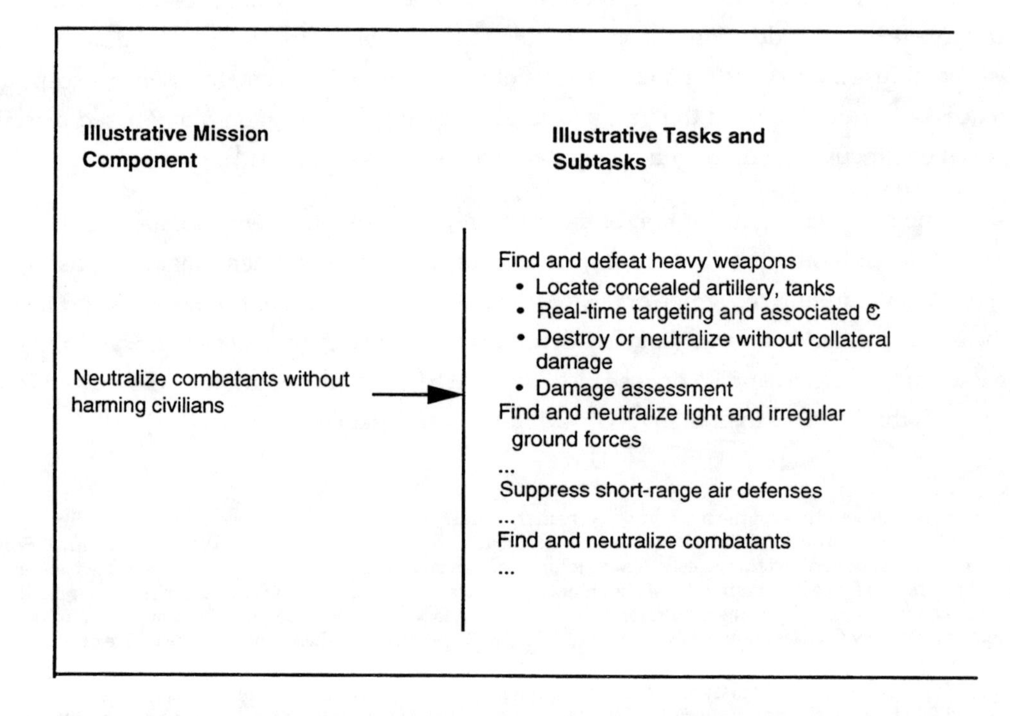

Figure 16—Decomposition into Tasks and Subtasks

Subtask / System	Locate concealed artillery and tanks	Real-time targeting and C3
Counter-battery radars and artillery	Low	Low
SOF detachments	Low	Low
Army Corps-level SLAR and ELINT	Low	Low
Scout and attack helicopters	Low	Low
JSTARS	Low	Low
...	...	...

Figure 17—Notional Scorecard for Evaluating Systems Against Subtasks

Concept Action Groups

If one has a sense of the tasks to be performed, there then exists the challenge of developing serious concepts for doing so—preferably competitive concepts using different weapon systems and different approaches. This work is inherently creative, although it is also focused by the problem to be solved and requires analytic minds and technical knowledge. Hence, we include it in this section on managerial methods. Colleague Glenn Kent has for some years championed an effective method, which he calls the "concept action group." Figure 18 illustrates how the "conceivers" function performed in a concept action group relates to other activities (see also Kent, Crawford, and Bonds, 1995). The concept action group brings together the "operators" (e.g., a CINC concerned about developing new capabilities for the task at hand), technologists, and analysts. Technologists describe the technical capabilities that can be brought to bear; the operators keep their minds on the task to be accomplished; the group as a whole conceives alternative concepts (Figure 19); and analysts study their relative virtues. Although this may seem unremarkable, it is in practice an unusual activity because people of a given type (e.g., operators, technologists, and so on) normally stick with their own and conduct disconnected studies. There is no published primer on how to proceed, but briefings are available and the experience base includes the 1995 summer study of the Air Force Scientific Advisory Board (the Vista study), although the unclassified summary of that study reads more like a straight

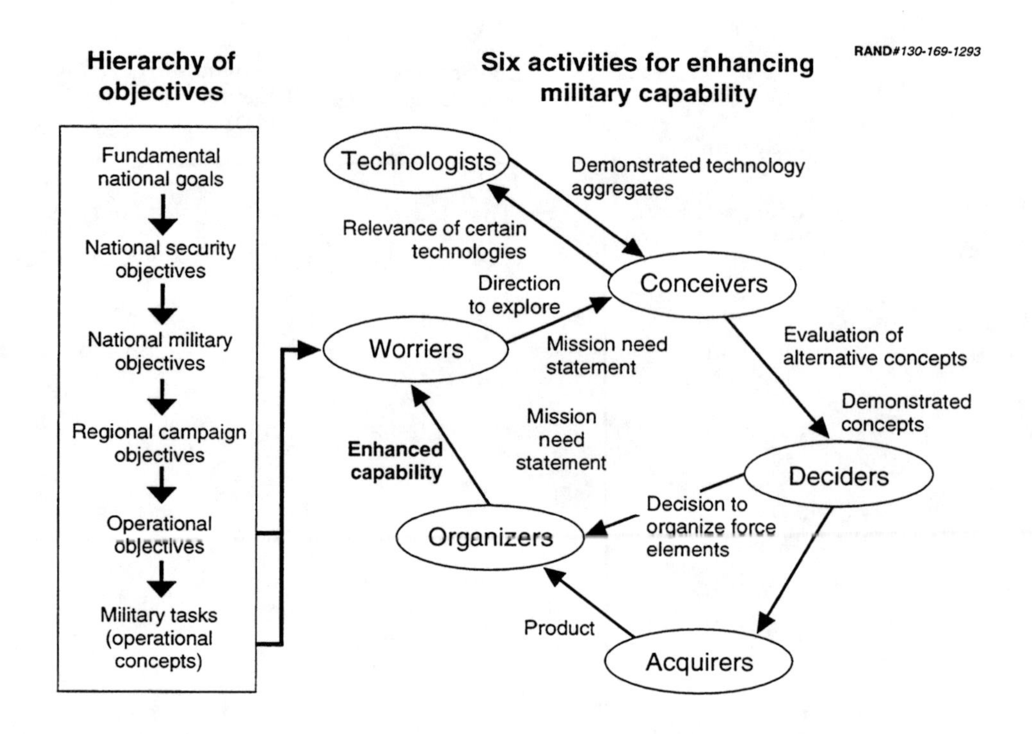

Figure 18—The Concept Action Group (Conceivers) in a Larger Context

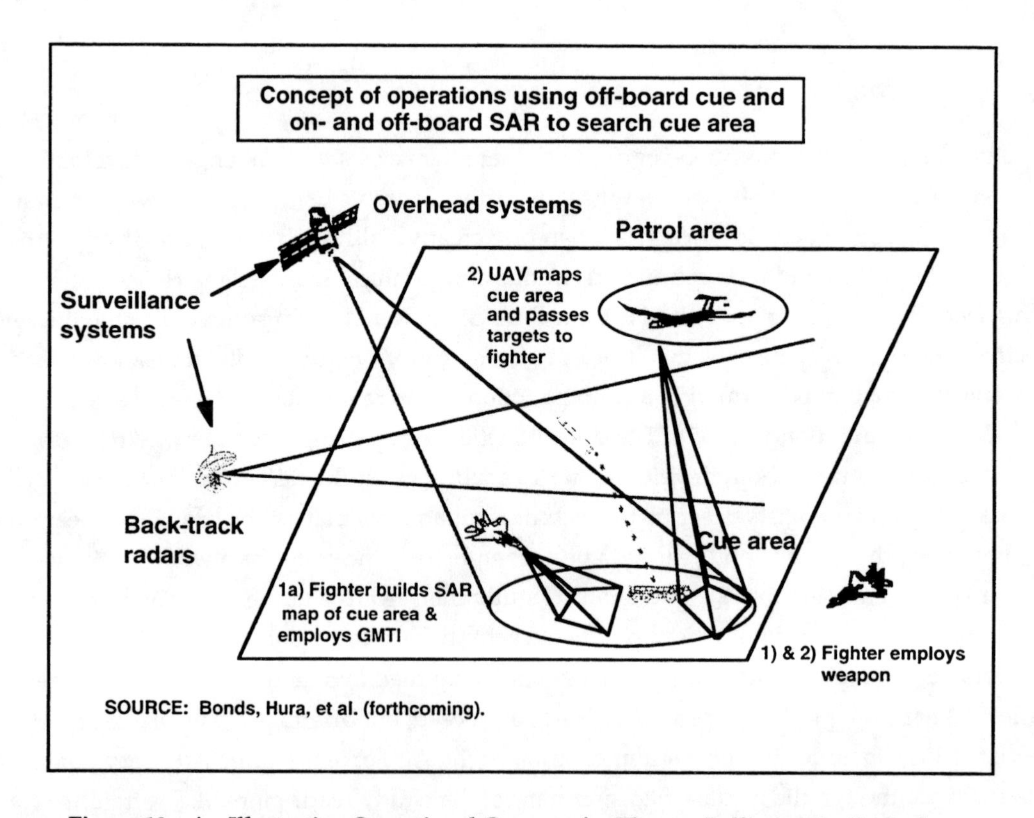

Figure 19—An Illustrative Operational Concept for Theater Ballistic Missile Defense

technology forecast. Also, a current RAND study for OSD directed by John Birkler is using concept action groups to study operations involving urban warfare.

Assessing Alternatives and Integrating: Moving to a Plan

Given a body of higher-level objectives and strategies, and given a body of alternative concepts for accomplishing them, the next class of activities involves assessment and moving toward a plan. What methods can be used for that purpose?

Threat-Based Planning (Scenario-Based Planning)

We mention threat-based planning only briefly here because it is so familiar (and was used for the BUR). This is the approach used since the early 1960s by OSD (Kaufmann, 1982; Davis, 1994c). It focuses planning around one or a very few partially defined scenarios (e.g., the Warsaw Pact invades the Central Region of NATO with 90 divisions, the attack beginning after 20 days of mobilization, with NATO starting its own mobilization on D-18). Ideally, the scenario reflects a greater-than-expected, but reasonably realistic, depiction of the threat. Building forces to deal with the threat results in the capability to handle a variety of other scenarios, regarded as lesser-included cases. The approach had a number of advantages, including an apparent simplicity. The case for the defense program could be explained easily to Congress and the public. Further, the threat scenarios could be used to help manage the DoD by establishing common cases against which Service programs would be measured. Although we do *not* believe threat-based planning with only one or a few point scenarios is an appropriate basic method for Air Force plans when thinking beyond the near term, we suggest a variant below called "stressful scenario sets," which should be informed by broader analysis.

Adaptive Planning (Capabilities-Based Planning)

Planning for adaptiveness (also called adaptive planning) is a method developed[13] in a series of studies for OSD and the Joint Staff. It includes what one of us (Davis) previously called capabilities-based planning, a term with too many conflicting meanings. On the one hand, planning for adaptiveness is simple in concept and consistent with a common-sense image of what military planning might be like. On the other hand, it is so much at odds with the more familiar threat-based approach and with the Joint Staff's traditional "deliberate planning system" as to appear very complex.

[13]See Davis (1994b) and an earlier study by Davis and Louis Finch cited there; see also Davis (unpublished), which describes the first phase of a major RAND project to implement many of the planning concepts. The study is sponsored by the oversight group for RAND's National Defense Research Institute (NDRI), which includes senior representatives from a number of OSD offices, the Joint Staff, and defense agencies.

Scenario-Space Contingency Analysis

One key element of the approach is "scenario-space contingency analysis," in which the goodness of a force posture as a whole (or the value of a particular improvement measure) is measured against a vast range of scenarios and scenario details, as suggested in Figure 20. This approach rejects utterly the notion that we can reliably forecast the contingencies in which future forces will be involved, much less the scenario details. Thus, this view of the world encourages development of forces that are highly flexible and force postures and operational plans that are quite adaptive, so that U.S. capabilities can be robust to variations of circumstance.

The basic ideas of adaptive planning are described elsewhere, but Figure 20 summarizes some key ideas. Instead of working with one or two Defense Planning Guidance "point scenarios," one wants to consider a *long* list of plausible contingencies, a list that includes politically incorrect cases and cases in which the U.S. military hopes that the United States never gets involved (e.g., a Chinese threat to Taiwan or a Russian threat to Lithuania). The list should be constructed by strategists, not regional specialists overly sanguine about their predictive capabilities or people worried about the political acceptability or economic consequences of scenarios.

This, however, is only the first step. A crucial second step involves transforming each of the broadly defined scenarios (e.g., China versus Taiwan) into a "scenario space." There is an entire space of scenarios for China versus Taiwan with different assumptions about the political-military context (e.g., allies and cause of crisis), military strategies, force levels, force and weapon

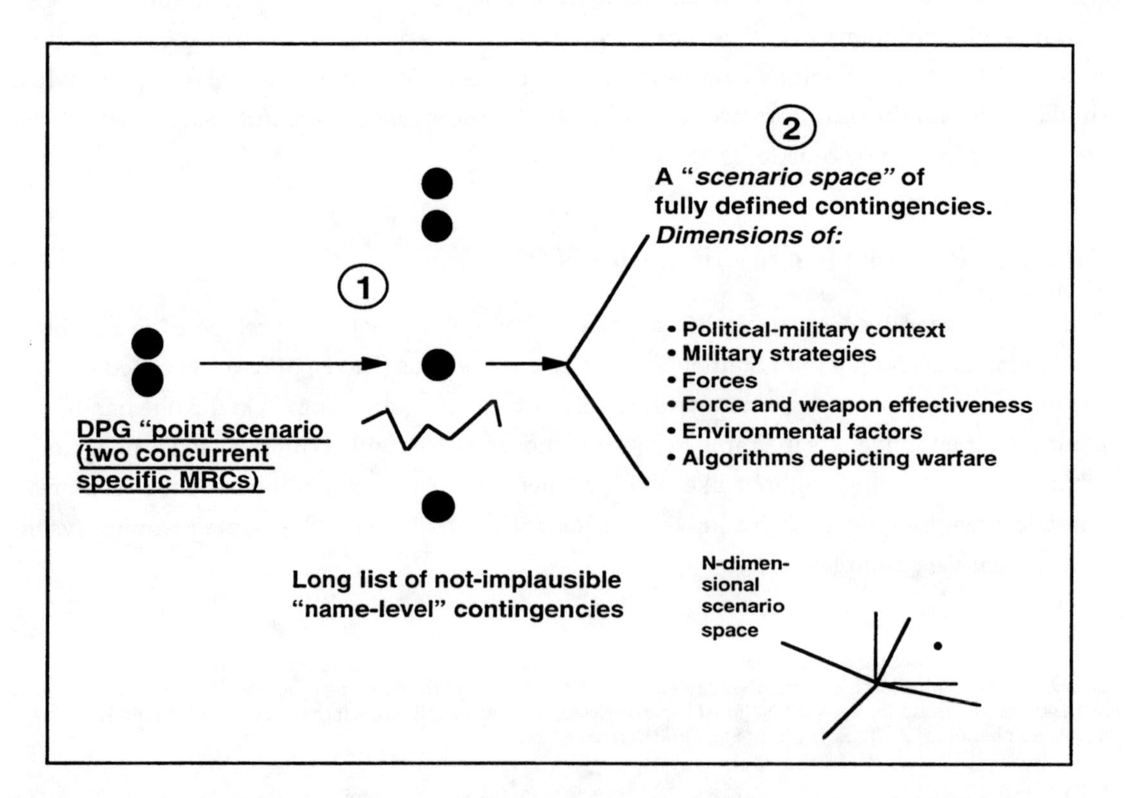

Figure 20—Expanding the Test Cases Against Which Forces Are Assessed: Planning for Adaptiveness

effectiveness, environment (e.g., weather), and even the algorithms used in military models. By conceiving the challenge in this way, one can "get out of the box" in recognizing the full range of challenges for which the United States may want to be prepared. In practice, of course, some detailed scenarios are far too hard to handle without spending enormous resources, and perhaps are so even then. Thus, a follow-on to scenario-space analysis is to make judgments about "how much is enough": On what portions of scenario space should we focus when building our future military posture, given considerations of national interests, scenario feasibility, expense, and so on? How does this vary with budget level?

As one illustration of what emerges analytically, Figure 21 shows some results of thousands of simulations to examine the feasibility of stopping a year-2010 Russian invasion of Poland well short of the Vistula as a function of how quickly NATO reacts (measured relative to D-Day), the effectiveness of tactical air (in kills per sortie),[14] and the extent to which sorties are suppressed (e.g., by chemical attacks on airfields, bad weather, or a fierce air-defense system requiring a lengthy SEAD campaign and more restricted air-to-ground operations). The idea in viewing such

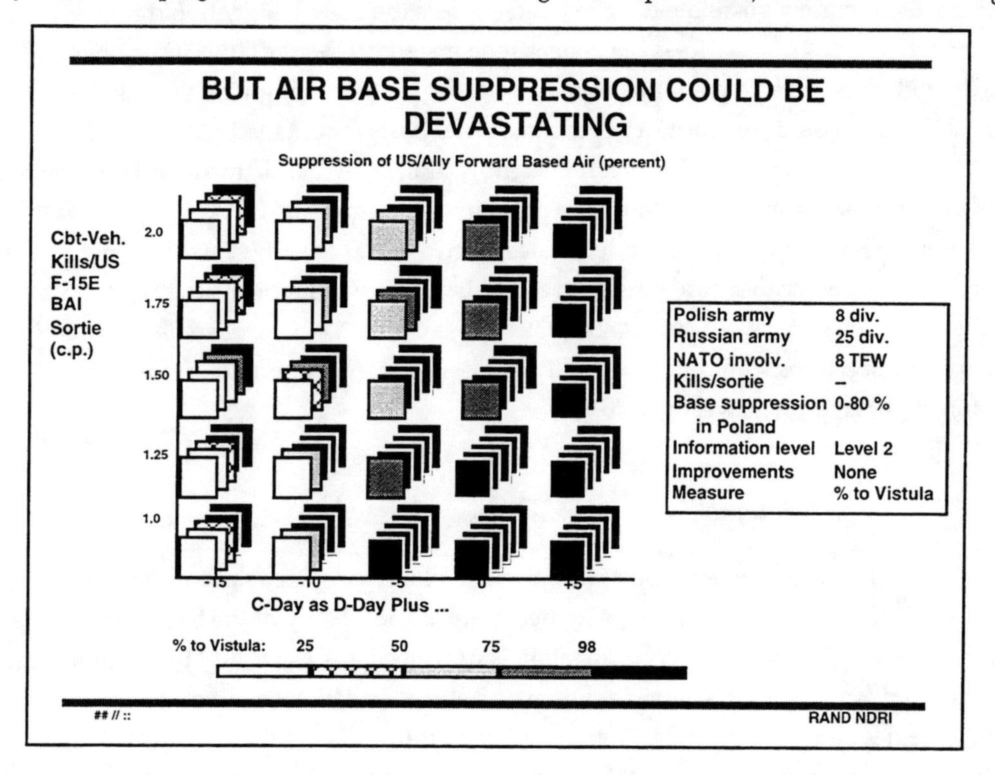

NOTE: Color of squares denotes the outcome of a simulation; white is "good"; black is "bad"; the into-the-paper axis is percentage of sorties suppressed, e.g., by air base attacks.

Figure 21—An Example of a Scenario-Space Depiction of Capabilities

---

[14]The y-axis gives the baseline value of the kills per sortie of an F-15E assumed for the particular case. If Level 2 information dominance is assumed, the effectiveness is raised another 25 percent. However, in the simulation, there are estimates of redundant targeting of targets and other frictional problems of imperfect C4I/ISR, which reduce the net effectiveness to some extent. Future analysis will consider aircraft with greater weapon loads and do a more fine-grained look at the effects of greater or poorer C4I/ISR, and faster or slower SEAD campaigns.

displays is to get a sense about for what portion of "scenario space" our forces should be quite adequate (white in the figure), quite poor (black), or somewhere in between. Force-improvement measures are chosen so as to push the envelope outward—i.e., to expand the region of scenario space for which capabilities would be adequate. Costs matter greatly, of course, in deciding how to expand coverage of scenario space.

Planning Scenarios for Management: "Stressful Scenario Sets"

It is probably inappropriate and impractical for the DoD itself to use sophisticated methods routinely, such as the scenario-space analysis described above. Instead, the DoD needs to use such methods in background studies, then to decide on planning scenarios that constitute appropriately stressing requirements for the Services and agencies to accomplish. There has been great resistance to the introduction of appropriately stressing scenarios, however, because DoD's scenarios have served so many purposes (declaratory strategy, arguments to Congress, and so on). Here, let us merely postulate the need and express the view that it should probably be possible to extract from scenario-space analysis the insights necessary to put together approximately three sizable planning scenarios per region (what might be thought of as MRCs or larger LRCs) that would constitute a good managerial basis for detailed planning. These three "cases" of a regional scenario might be, for example, Iraq reinvades Kuwait and goes on to Saudi Arabia under three sets of circumstances: a set akin to the current DPG scenario; a set that posits very late-in-crisis decisions by the United States, such as C=D or C=D+2, which might necessitate a degree of forced entry to secure ports and airfields; and a set that posits the United States having to reinvade to recover Saudi Arabia and Kuwait after a period of some months. Other features of the scenarios would involve ballistic missile defense, operations in a chemical environment, counteroffensives, and so on.

Choice-Oriented Systems Analysis

There is no well-defined methodology for making the kinds of system and mission-level tradeoffs that are so critically needed in the years immediately ahead. Many of the techniques introduced in the 1960s still apply, but more sophistication is now needed—especially for planning against diverse scenarios and scenario details as discussed above. A key element of the adaptive planning we describe above should probably be something analogous to work RAND recently conducted for the Dutch government (Hillestad et al., forthcoming). This involves spreadsheet-level models, a range of challenges against which to measure things (e.g., the Stressful Scenario Sets mentioned above and, where possible, a more fulsome scenario space), a large set of potential "tactics" and "strategies" (individual force-improvement measures and packages thereof that can be associated with a concept, such as focusing on long-range strike by manned aircraft). The result is the ability to examine interactively the relative cost-effectiveness of alternatives while changing subjective assessments about the relative importance of different classes of warfighting scenario, assumptions about the forces themselves, and so on. This work is still in its infancy but is being pursued vigorously by Hillestad in an Air Force–sponsored project led by colleague Natalie Crawford.[15]

Strategic Portfolio Analysis

Although we do not discuss the issue here in any detail because the methods are still being developed, a key element of RAND's approach to adaptive planning is something we call strategic portfolio analysis. The problem this addresses is that force planning can no longer be built around contingency analysis alone. Instead, the challenges of environment shaping and strategic adaptiveness through hedging (i.e., having capabilities that could be expanded readily in response to major changes in strategic environment, such as development by China of a highly effective Navy and force-projection capabilities against Taiwan) may be just as important in dictating future force structure as the needs assessed in examining future warfighting (Davis, 1996).

To deal with this reality, RAND is developing portfolio methods to help policymakers make resource allocation decisions across categories of mission that are essentially "apples and oranges" (Figure 22). Some of this is building on work for the Joint Staff's J-5, which has been concerned about resource-allocation strategies for improving the effectiveness of "presence" in all its many forms (see Kugler, unpublished, and Winnefeld, unpublished).

For the Air Force, a "portfolio of activities" might be characterized by different criteria, such as the ability to meet mission requirements, capacity to shape a military or even a political environment (e.g., with "virtual presence"), and the ease with which systems can be used as a base for more-advanced technological developments, if needed. The basic point here is to go beyond only battle engagements and threat-based planning as a framework for resource allocation.

Strategically Adaptive Planning

Let us mention briefly an emerging method for strategic planners to use in establishing an adaptive plan to be applied over a period of years in response to events. This approach is an outgrowth of work alluded to earlier on exploratory analysis methods in which one confronts massive uncertainty rather than focusing on best estimates (Figures 20 and 21; see also Bankes, 1993, and Bankes and Gillogly, 1994). However, it also reflects the influence of the last 15 years' work on complexity theory and, more particularly, *complex adaptive systems*. As illustrated in a recent study on policy choices regarding global warming, Lempert, Schlesinger, and Bankes (forthcoming) demonstrate how, in the face of large uncertainty about the consequences of global

[15]This approach seems to capture the features called for by Maj Mace Carpenter (USAF) in a short presentation on "effects-based planning" in a RAND–Air Force workshop (Bordeaux and Ochmanek, unpublished). Carpenter correctly noted that much more emphasis should be placed on tradeoff analysis at the level of competing ways to accomplish operational objectives and tasks.

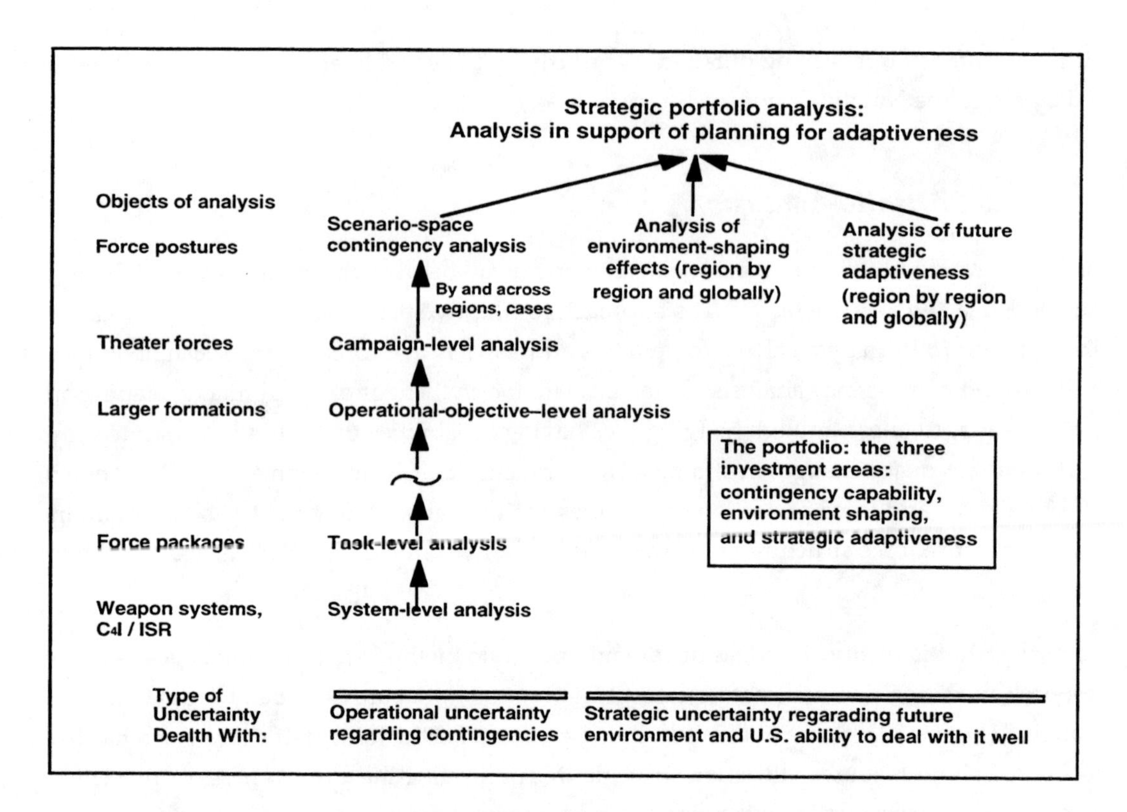

Figure 22—Contributors to Analysis of Adaptiveness

warming, the cost of remediation, and other factors, a wise strategy is one that eschews either of the emotionally argued pure strategies (roughly, "major remediation steps now" versus "wait and see") and instead involves adjusting the pace of remediation efforts according to a very simple algorithm as data emerge over time (Figure 23). This strategy dominates the pure strategies for a very large portion of the uncertainty space (Figure 24). [16]

The approach taken in this study is, we believe, remarkably general in its applicability. It is consistent intellectually with a long stream of management and decision studies illustrating the power of relatively simple heuristics. It is a strong method for analyzing alternatives, for integrating to find a more robust strategy, and for specifying how to monitor and adapt over time. Much more work needs to be done to perfect these methods, however. In practice, they require analytic depth, theory, models, and sophisticated computer tools.

Affordability Analysis

Although we shall not discuss it in any depth here, affordability analysis should be a key element of planning, especially when moving toward development of integrated programs and a detailed

[16]To some extent, one can consider this work as consistent with the spirit and insights of Lindblom's efforts several decades ago, which were not, however, accompanied by analytical methods such as are now possible.

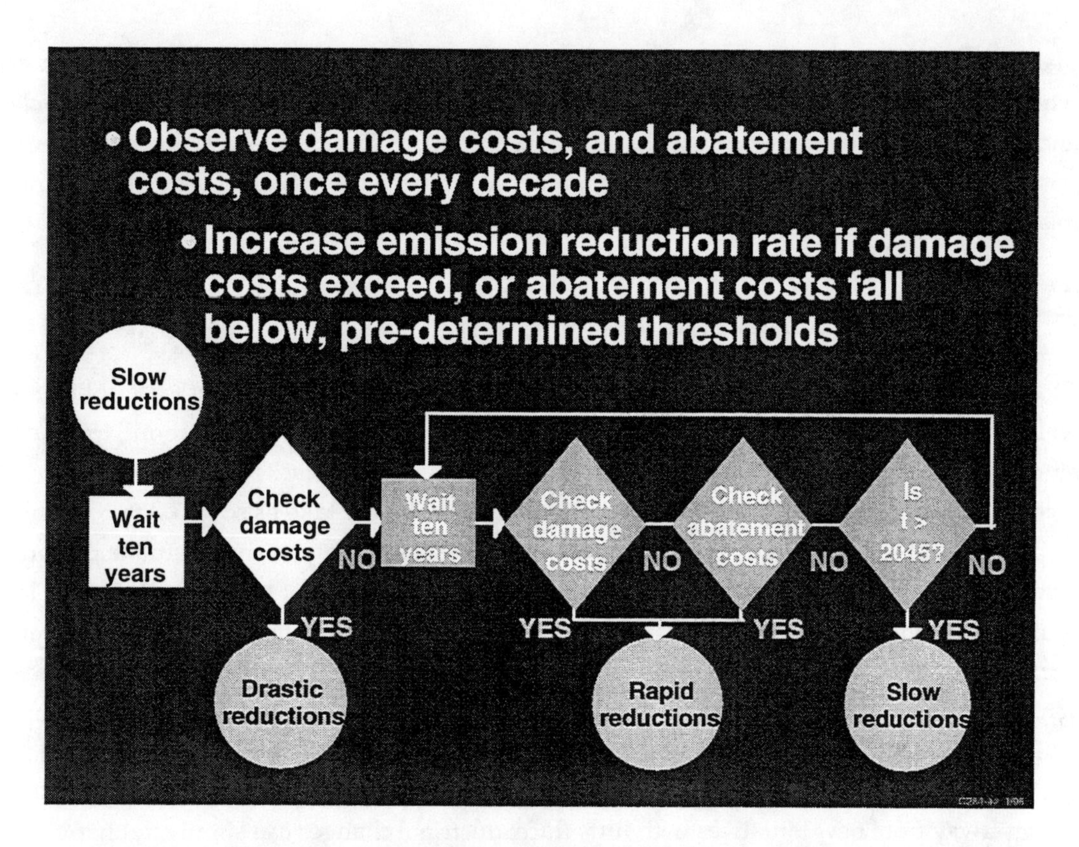

Figure 23—A Simple Algorithm for Adaptation over Time

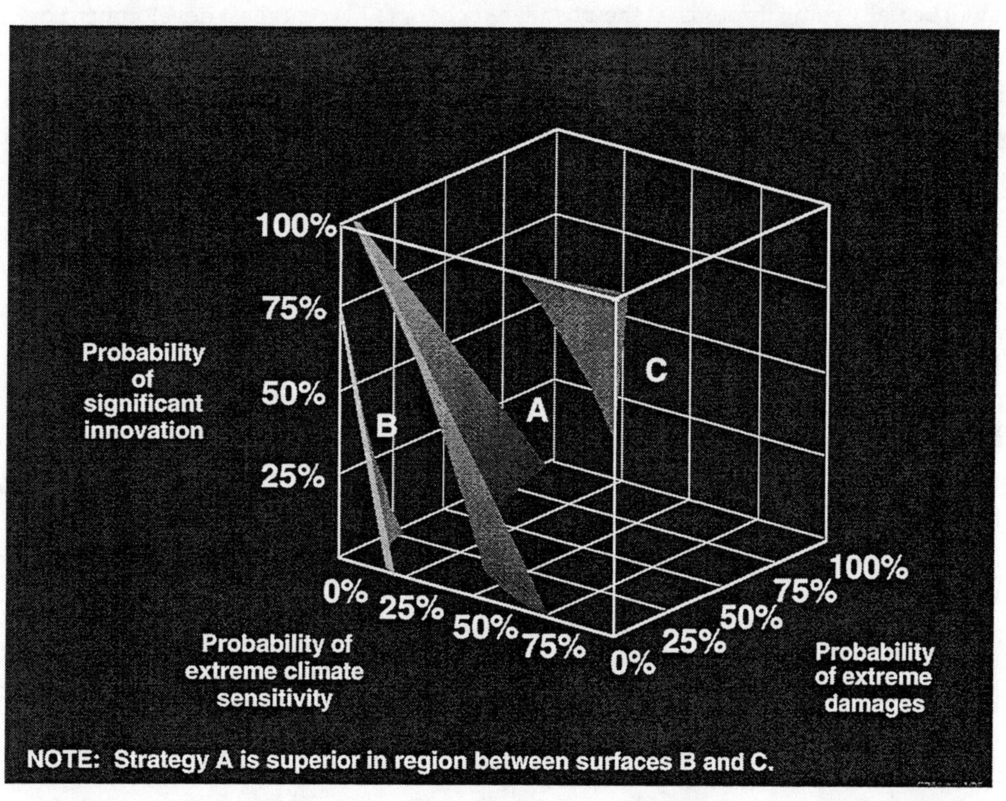

Figure 24—Regions of Uncertainty Space in Which the Alternative Strategies Dominate

plan over time. Affordability analysis takes a life-cycle view of costing, but also keeps track of when the bills will come due in relation to the many other bills that arise, some of them cyclically and some of them routinely. Affordability analysis may not sound "strategic" to visionaries, but it is often a major determinant in actual planning. For a good discussion of affordability analysis in an Air Force context, see Stanley (1994).

Organizational Viability Analysis

To end this section, we suggest a method that ought to exist, but does not—at least not in explicit, well-defined terms for people to study and employ. The issue here is that, in trying to choose among competing operational concepts for accomplishing various tasks and operational objectives, it is quite possible that quantitative analysis will only take us so far. After all, the protagonists of the competing approaches can all do the same analysis and postulate performance levels and doctrinal procedures sufficient to make their concepts look good. Further, in at least some cases, competitive approaches may indeed be equally credible in technical and theoretical terms. What then? In our view, a key issue is which organization or team of organizations will actually make the necessary changes in program and doctrine to accomplish the concept for which they seek money. It is one thing for bright young officers to hypothesize changes; it is quite another for the necessary changes to occur and stick. Reprogramming actions can shift money away from new initiatives with little trace; doctrinal changes can simply not happen; and so on. All of this suggests that a *major* factor in integrative planning must be an assessment of who will actually come through on the postulated changes. We suggest development of something that might be called organizational viability analysis, which would be a more formalized effort to force proposing organizations to define how they would accomplish what they are proposing, and what assurances would exist about protecting their priority. There are examples in past planning of this type of thing, but it has not been formalized. We believe it should be.

If there were a formalized organizational viability analysis, we believe the Air Force would be strongly competitive in some respects and less credible in others. On the one hand, the Air Force has unsurpassed expertise in certain types of command and control, including use of information from satellites, aircraft, and ground facilities. Also, Air Force systems were highly effective in attacking both facilities and armored vehicles in Desert Storm: The Air Force had procured the necessary weapons and developed the necessary doctrine. On the other hand, the Air Force has not moved as rapidly as critics would prefer in procuring more advanced munitions, in equipping the B-2 with the avionics and weapons that would make it an effective weapon system in the instance of a surprise attack, or in developing and fielding UAVs.

5. Comparison of Methods

Factors Influencing Choice of Methods

Having surveyed strategic planning methods, the next challenge is to provide some guidance as to when a given method might be useful for the Air Force. To address this, let us first consider what some of the key factors might be. They would seem to be the following:

- *Degree of creativity sought.* Is one trying to "get out of the box" or to do more routine, albeit potentially difficult, problem-solving?
- *Time scale.* Is one looking out into the near future (a few years), the mid-term (perhaps 3–7 years), or the long term (perhaps 8–15), or the very long term (perhaps 20–40 years)?
- *Type of impact sought.* Possibilities here include:
    - — Seeking new insights (e.g., discovering and appreciating vulnerabilities, advantages, implications of technological change)
    - — Exploration (e.g., of plausible new doctrinal methods exploiting new weapon systems)
    - — Seeking knowledge (e.g., of what to expect in terms of action-reaction cycles)
    - — Sensitizing participants to a problem or for the need for organizational and doctrinal change (e.g., to the likely need for U.S. forces actually to operate in an WMD environment, most likely chemical)
    - — Teaching new paradigms (e.g., demonstrating the benefits of using new doctrine).
- *The participants.* In thinking about the planning exercise itself, who will be participating: national-level officials, the Chief of Staff, other general officers, etc.?
- *The consumer.* When the planning exercise is completed, what is the product and to whom does the product go? The exercise may itself be the product in some cases, but in most cases the consumer of strategic planning should be the most senior officers of the Air Force.[1]

Comparisons

It is inherently difficult to compare the various methods because so many of them are flexible, with the result that what they are depends on who uses them and in what context. Nonetheless, Table 1 gives a subjective summary.

[1]It is well accepted among strategic planners that the value of the work they do is exceedingly dependent upon the top executives of the organization being directly interested and involved. Strategic planning conducted by staffs for staffs is often a virtual waste of time.

Table 1

Comparison of Planning Methods

Method	Creativity	Time Scale	Type Impact	Participants	Consumer
Uncertainty-Sensitive Planning	Very strong	Short, Medium	New insights, exploration, strategy	Staffs	Policymakers
Assumption-Based Planning	Very strong	Short, Medium, Long	Review and improvement of plans	Staffs	Policymakers
Alternative Futures (Alternative Scenarios)	Very strong	Short, Medium	New insights, exploration, strategy	Staffs	Policymakers
Future-of-Warfare Gaming	Very strong	Short, Medium, Long	New insights, exploration, concepts	Staffs	All levels
Day-After Gaming	Very Strong	Short, Medium	New insights, sensitization	Policymakers and staffs	All levels
Objectives-Based Planning	Relatively analytic	Short, Medium	Knowledge, structure, discipline	Staffs	Senior leaders and managers
Mission-Pull Planning	Relatively analytic	Long	Structure, discipline, sense of priorities	Staffs	Senior leaders and managers
Concept Action Groups	Engineering-and opera-tions-level creativity	Short, Medium, Long	Implementable concepts, insights	Staff and general-officer operators	Senior leaders and managers
Threat-Based Planning	Low	Short	Has been useful in *implement-ing* strategy	Analytic staffs	Multiple levels of organi-zation
Planning for Adaptiveness (Capabilities-Based Planning)	Very strong for going beyond standard scenarios	Short, medium	Insights about Achilles' heels and high-leverage improvement measures	Analytic staffs	Policymakers
Stressful Scenario Sets	Simplified way of going beyond standard scenarios	Short, medium	Management device for focusing activities	Staffs	Multiple layers within acquisition
Strategic Portfolio Analysis	Resource allocation across types of mission	Short, medium, long	Decision support for inherently subjective decisions	Staffs and policymakers	Policymakers
Strategic Adaptation in Complex Adaptive Systems	Requires analytical creativity and abstraction	Medium, long	Decision support for strategic adaptation	Staffs and policymakers	Policymakers

Table 2 arrays the various methods against the planning activities discussed earlier, with more checks indicating more usefulness. As in all such depictions, some of the assessments are quite subjective.

Table 2

Methods for Different Planning Activities (more checks denote more usefulness)

Method	1. Recognizing and conceptualizing challenges and objectives	2. Conceptualizing and defining strategies	3. Analysis: assessing the alternatives	4. Integrating, choosing, refining, and building a robust plan	5.–6. Implementing, explaining, monitoring, and adapting the plan
Top-Level Structuring	√√√				
Uncertainty-Sensitive Planning	√√√	√√√			
Alternative Futures and Technology Forecasts	√√	√√			
Out-of-the-Box Gaming	√√√	√√			√√
Assumption-Based Planning		√√√		√√√	
Objectives-Based Planning	√√√	√√√		√√	√√√
Concept Action Groups		√√√			
Systems Analysis of Contingency Capability			√√√	√√	
Stressful Scenario Sets			√	√√√	√√
Strategic Portfolio Analysis				√√√	√
Strategic Adaptation for Complex Adaptive Systems			√√√	√√√	√√√
Organizational Viability Analysis			√√√	√√√	
Affordability Analysis			√	√√	√√√

Finally, Table 3 summarizes the material in a way that relates methods to the products of the formal planning process. Again, our observations are subjective. Table 3 shows three blocks of methods as discussed earlier. The first applies to helping establish high-level objectives and strategies, the second to structuring and managing (including the development of alternative operational concepts to accomplish tasks), the third to analysis and integration.

Conclusions and Recommendations

Our basic recommendation is that the Air Force view strategic planning methods as a package with different methods for different purposes (Table 2). We see right-brain exercises in creative divergent thinking as crucial in the formulation of strategy (e.g., Uncertainty Sensitive Planning). We see a different set of methods (Objective-Based Planning with some features from Mission-Pull Planning) as essential for providing structure and managerial priorities. We see still other methods (e.g., the Concept Action Groups) as well suited to turning broad strategies and priorities into down-to-earth system concepts that can be used to define acquisition strategies, programs, doctrinal change, and so on. We see Adaptive Planning (capabilities-based planning) as the method of choice for making resource-allocation decisions across programs. This must include systems analysis and Strategic Portfolio Analysis as a mechanism for explicitly integrating, albeit subjectively, across major missions such as environment shaping.

In concluding this study we remind the reader that strategic planning is an art and craft, not a science. To be sure, we can identify the major activities and suggest a broad range of methods for conducting them. Ultimately, however, the quality of the results will be no better than the people involved and no better than the circumstances will allow. The methods we have described have proven valuable in past work, but they are only methods. They can readily be undercut by constraints limiting the range of futures or options to be considered, or by giving control to people overly wedded to present circumstances and conventional wisdom regarding the future. Ironically, one of the problems that arises is that participants may find themselves subject to peer pressure regarding what forward-thinking people are "supposed" to believe about the future. Rather than being tied to the past, then, they may be caught up in a fad (we recall interest in nuclear airplanes during the 1950s and suspect that some of the current panacea-seeking enthusiasm for casualty-free long-range precision strike will be dampened when people again consider the nature of infantry-intensive warfare in mountains, jungles, and urban settings, and when countermeasures appear). While these observations may seem obvious and unnecessary, they are not so easy to heed. It is worth remembering that many long-term planners 20 years ago were subject to conventional-wisdom pressures to the effect that long-range penetrating bombers would obviously be obsolete, that carrier battle groups would obviously be highly vulnerable, that ground forces would continue to dominate warfare, and that, of course, the Soviet Union would long be a superpower competitor. The humility gained from remembering this should perhaps encourage planners in 1996 to be cautious in assuming they can predict confidently the nature of future warfare or the identity and nature of our principal future adversaries.

Table 3

A Composite Approach to Force Planning: Products Sought and Methods to Be Used

Product	Methods Useful in Developing Product	Comments
National Security Strategy (NSS) and National Military Strategy (NMS)	Top-Level Structuring	Establishes a coherent framework tied to fundamental interests and top-level objectives.
	Uncertainty-Sensitive Planning (USP) (including looks at alternative strategic environments, budget levels, and views of national interests)	Premium is on open-minded divergent thinking, followed by synthesis. Output of creative and analytic exercises may or may not be clear-cut decisions, but will include insights affecting plan-level decisions.
	Alternative Futures and Technology Forecasts	Focus is on bringing out alternative images of the future with respect to both the external environment and the national strategy, and with respect to technology.
	Out-of-the Box Gaming	Purposes include thinking the unthinkable and conceiving new strategies.
	Assumption-Based Planning	Encourages creative strategy by critiquing a baseline—by identifying fundamentally important but implicit assumptions that could fail.
Joint Missions and Operational Objectives	All of above techniques	
	Objective-Based Planning (strategies to tasks) conducted for a wide range of circumstances	Premium is on top-down structured analysis. Output is a taxonomy of well-defined functions to be accomplished, motivated by national strategy and its priorities.
Joint Tasks	Objective-Based Planning	Premium is on translating abstract functions into concrete tasks suitable for practical management.
Operational Concepts	Concept Action Groups Comparative systems analysis	Premium is on creative but pragmatic work producing concrete system concepts for accomplishing the various tasks and missions, followed by objective tradeoff analyses to help choose among competitive concepts.
Body of Analysis and Tentative Choices of Approach	Program analysis (especially tradeoff studies and marginal analysis with attention to diminishing-return curves)	Objective is to translate operational concepts into programs for procurement, doctrinal change, training, and so on, and, again, to analyze alternatives.

Table 3 —continued

Product	Methods Useful in Developing Product	Comments
Defense Program and Posture (and, within it, the Air Force program)	Adaptive Planning (which includes capabilities-based planning) using Strategic Portfolio Analysis	Objective is to assess programs and postures, for different budget levels, against a broad range of future contingencies (scenario-space analysis) and against needs to influence the strategic environment and be prepared for strategic adaptation.
	Strategic Adaptation in Complex Adaptive Systems	Objective is to follow a hedged approach initially and to adapt in particular ways in response to specified measures of need.
	Stressful Scenario Sets	Purpose is to simplify expression of requirements for management of programs and other activities.
	Assumption-Based Planning	Purpose is to review and amend plans to better cope with uncertainty.
	Affordability Analysis	Purposes include providing a life-cycle view of costs, timing major investments to avoid budgetary shocks or temporary losses of capability, and providing reality checks on what can actually be fit into a program with finite resources.
	Organizational-Viability Analysis	Purpose is to assess the competing organizations' ability and willingness to make the changes and investments needed for success of a program (undeveloped).

NOTE: Although the nominal planning process may be linear, the reality is distinctly nonlinear, with many activities in parallel and with many feedback loops. Thus, this table should not be read to imply that a method higher in the table should necessarily be used before a method listed below it. Neither is national military strategy determined in a vacuum. Instead, it reflects a sense of the consequences for operational-level objectives and the defense program. The clear top-down depiction of planning is more realistic in describing results than in describing intellectual processes.

Appendix

Common Pitfalls of Strategic Planning

Based on our personal experiences and our knowledge of the relevant literature, we would caution Air Force planners on the following common dangers and pitfalls:

- Allowing strategic planning to degenerate into what amounts to incremental adjustments of a preexisting strategy without full discussion of the strategy itself and alternatives. This becomes a major pitfall as soon as one begins to routinize the planning activities or when "everyone knows" that many of the tough issues are out of bounds for discussion.
- Focusing on "forecasts" of any sort, and focusing solely on so-called best-estimate predictions in particular. The problem here is that forecasts are usually mere projections of what already exists or is in process of coming into existence. The real world, by contrast, often develops quite differently.
- Believing *either* that the future is unknowable *or* that uncertainties are modest. Realistically, many features of the future are predictable, and many others should be predicted only with considerable humility.
- Taking a purely reactive stance. It is important for planners to develop hedges and contingency plans, but they should also be seeking to influence what the future will be.
- Believing that the future can be controlled. The flip side of the previous pitfall is believing that one can not only "influence," but positively "control" the future. That cybernetic ambition is sensible when dealing with simple systems, but not systems as complex and nonlinear as the international environment. The paradigm of complexity theory is more apt.
- Formulating futuristic scenarios in such "pure" form as to make most of them patently absurd. The danger here is that some features of "crazy" scenarios may turn out to be realistic. For example, it may be "crazy" to spend much time contemplating a world in which there is massive and widespread terrorism on the scale of the World Trade Center disaster, but ignoring the possibility that terrorism might become a key element of conflict—e.g., being used to influence American decisions and actions during critical periods of time during crisis—would be most unwise. Similarly, while the emergence of a true global peer competitor may be crazy within the next 10–15 years, it is by no means absurd to imagine that one or more nations might develop capabilities that would severely challenge us in limited regions and in limited ways (e.g., denying us the ability to operate surface naval vessels with impunity within 1,000 km of a crisis area).
- "Believing" blue-sky concepts and speculations without relatively detailed follow-up analysis. Many participants in creative war games or related exercises become convinced that some of the postulated capabilities and tactics are obviously feasible and desirable.

Further scrutiny, however, will often demonstrate that they have fatal flaws (e.g., astronomical costs, violations of the laws of physics, vulnerability to countermeasures, and so on).

- Allowing political judgments to rule out possibilities, as in "the American people would never go for it" or "there is no way that the ally in question would permit this." This is a difficult subject, because there *are* limits on what it is worthwhile to consider. However, many of the people eager to offer such political judgments are much better at describing today's attitudes than at predicting the future. Attitudes change as circumstances change.
- Focusing exclusively on military and technological considerations without regard to political-military considerations. The flip side of the preceding item is that U.S. strategic planning should avoid being too unilateralist: In the real world, national security strategy and national military strategy will always depend *critically* upon allies. Indeed, it is strongly in the U.S. interest that they should.
- Confusing the role of "pure" futuristic scenarios and hybrids. Analytically, it is often useful to describe pure cases, while recognizing that the true future will almost surely be a combination case. However, when attempting to help leaders make actual decisions, the most useful options are likely to be ones that combine features of various pure strategies.

Bibliography

Arquilla, John, and Paul K. Davis, *Modeling Decisionmaking of Potential Proliferators as Part of Developing Counterproliferation Strategies*, Santa Monica, Calif.: RAND, MR-467, 1994.

Aspin, Les, *Report on the Bottom-Up Review*, Department of Defense, October 1993.

Bankes, Steven C., "Exploratory Modeling for Policy Analysis," *Operations Research*, Vol. 41, No. 3, 1993, pp. 435–449. Also available as a RAND reprint: RP-211.

Bankes, Steven C., and James Gillogly, "Exploratory Modeling: Search Through Spaces of Computational Experiments," *Proceedings of the Third Annual Conference on Evolutionary Programming*, 1994. Also available as a RAND reprint: RP-345.

Bennett, Bruce W., Sam Gardiner, and Dan Fox, "Not Merely Planning for the Last War," in Davis (1994d), 1994, pp. 477–514.

Bonds, Tim, Myron Hura, Keith Henry, Jeff Hagen, Dave Frelinger, Brian Nichiporuk, Bill Stanley, Dan Norton, and John Bordeaux, *An Assessment Framework for Assessing Military Capability* (U), Santa Monica, Calif.: RAND, DB-166-AF, forthcoming. Classified publication, not for public release.

Bordeaux, John and David Ochmanek, *Approaches to Force Planning: Results of a Workshop*, Santa Monica, Calif.: RAND, unpublished.

Bracken, Paul J., *Strategic Planning for National Security: Lessons from Business Experience*, Santa Monica, Calif.: RAND, N-3005-DAG/USDP, 1990.

Carpenter, Major Mace, "Effects-Based Planning," workshop presentation at RAND, November, 1995.

Cetron, Marvin, "An American Renaissance in the Year 2000," in Marvin Cetron, *American Renaissance*, 2nd ed., New York: St. Martin's Press, 1994.

Davis, Paul K., *National Security Planning in an Era of Uncertainty*, Santa Monica, Calif.: RAND, P-7538, 1989.

_____, "Protecting the Great Transition," in Davis (1994d), 1994a, pp. 135–164.

_____, "Institutionalizing Planning for Adaptiveness," in Davis (1994d), 1994b, pp. 73–100.

_____, "Planning Under Uncertainty Then and Now: Paradigms Lost and Paradigms Emerging," in Davis (1994d), 1994c, pp. 15–57.

_____, ed., *New Challenges for Defense Planning: Rethinking How Much Is Enough,* Santa Monica, Calif.: RAND, MR-400-RC, 1994d.

_____, *Planning Future Forces: First-Phase Report*, Santa Monica, Calif.: RAND, unpublished.

Davis, Paul K., and John Arquilla, *Deterring or Coercing Opponents in Crisis: Lessons from the War with Saddam Hussein*, Santa Monica, Calif.: RAND, R-4111-JS, 1991.

Davis, Paul K., and Zalmay Khalilzad, *When Next with North Korea*, Santa Monica, Calif.: RAND, unpublished.

Dewar, James A., and Morlie Levin, *Assumption-Based Planning for Army 21*, Santa Monica, Calif.: RAND, R-4172-A, 1992.

Dewar, James, Carl Builder, W. Michael Hix, and Morlie Levin, *Assumption-Based Planning: A Planning Tool for Very Uncertain Times*, Santa Monica, Calif.: RAND, MR-114-A, 1993.

Donaldson, Gordon, and Jay W. Lorsch, *Decision Making at the Top: The Shaping of Strategic Direction*, New York: Basic Books, 1983.

Finn, Brian, *PATH Games: a Decision Making Tool*, Harold Rosenbaum Associates, Defense Nuclear Agency, DNA-TR-88-31-AP, 1988.

Hillestad, Richard, Warren Walker, Manuel Carrillo, Joseph Bolten, Patricia Twaalfhoven, and Odette Van de Riet, *FORWARD: Freight Options for Road, Water, and Rail for the Dutch*, Santa Monica, Calif.: RAND, MR-736-EAC/VW, forthcoming.

Hinkle, Wade, and Clark Murdock, "Long-Range Defense Planning," briefing slides prepared by OUSD(P)/Policy Planning, March 2, 1994.

Kahn, Herman, and Irwin Mann, *Ten Common Pitfalls of Systems Analysis*, Santa Monica, Calif.: RAND, RM-1937, 1957.

Kaufmann, William, *Planning Conventional Forces, 1950–1980*, Washington, D.C. : Brookings Institution, 1982.

Kent, Glenn A., *Providing New Capabilities to Meet Future Needs*, Santa Monica, Calif.: RAND, unpublished.

Kent, Glenn, and William Simons, "Objectives-Based Planning," in Davis, (1994d), 1994, pp. 59–71.

Kent, Glenn, Natalie Crawford, and Tim Bonds, *Concept Development: The Elixir for the Future*, Santa Monica, Calif.: RAND, unpublished.

Khalilzad, Zalmay, *From Containment to Global Leadership? American and the World After the Cold War*, Santa Monica, Calif.: RAND, MR-525-AF, 1995.

Klein, Gary A., Judith Orasanu, Roberta Calderwood, and Caroline E. Zsambok, eds., *Decision Making in Action: Models and Methods*, Norwood, N.J.: Ablex Publishing Corp., 1993.

Kleindorfer, Paul R., Howard C. Kunreuther, and Paul J.H. Shoemaker, *Decision Sciences: An Integrative Perspective*, New York: Cambridge University Press, 1993.

Kugler, Richard, *Toward a Dangerous World: U.S. National Security Strategy for the Coming Turbulence*, Santa Monica, Calif.: RAND, MR-485-JS, 1995.

_____, *Whither the U.S. Military Presence in Europe? A Strategic Portfolio Analysis of Emerging Requirements and Priorities Using RAND's New Programs-to-Objectives Methodology*, Santa Monica, Calif.: RAND, unpublished.

Lempert, Robert, Michael E. Schlesinger, and Steve C. Bankes, "When We Don't Know the Costs or the Benefits: Adaptive Strategies for Abating Climate Change," *Climatic Change*, forthcoming.

Levin, Norman, ed., *Prisms and Policy: U.S. Security Strategy After the Cold War*, Santa Monica, Calif.: RAND, 1994.

Lindblom, Charles, "On the Science of Muddling Through," *Public Administration Review*, Spring, 1959.

_____, *The Policy-Making Process*, Prentice-Hall, Englewood Cliffs, N.J.: 1968.

March, James, *A Primer on Decision Making: How Decisions Happen*, New York: Free Press, 1994.

March, James, and Herbert Simon, *Organizations*, New York: Wiley, 1958.

Mintzberg, Henry, *The Rise and Fall of Strategic Planning*, The Free Press, New York, 1994.

Millot, Marc Dean, Roger Molander, and Peter A. Wilson, *"The Day After. . ." Study: Nuclear Proliferation in the Post–Cold War World*, Vol. I, *Summary Report*, Santa Monica, Calif.: RAND, MR-266-AF, 1993.

_____, *"The Day After. . ." Study: Nuclear Proliferation in the Post–Cold War World:* Vol. II, *Main Report*, Santa Monica, Calif.: RAND, MR-253-AF, 1993.

_____, *"The Day After. . ." Study: Nuclear Proliferation in the Post–Cold War World:* Vol. III, *Exercise Materials*, Santa Monica, Calif.: RAND, MR-267-AF, 1993.

Mintzberg, Henry, *The Rise and Fall of Strategic Planning*, New York: The Free Press, 1994.

Murdock, Clark, "Mission-Pull and Long-Range Planning," *Joint Force Quarterly*, Fall, 1994.

_____, "Thinking Long-Range: About the Next Air Force . . . and the Air Force After Next," briefing slides, February 14, 1995.

Ochmanek, David A., and Stephen T. Hosmer, with John Bordeaux, *A Context for Defense Planning*, Santa Monica, Calif.: RAND, unpublished.

Peters, Thomas J. and Robert H. Waterman, Jr., *In Search of Excellence: Lessons from America's Best-Run Companies*, New York: Harper&Row, 1982.

Pirnie, Bruce, and Sam Gardiner, *An Objectives-Based Approach to Military Campaign Analysis*, Santa Monica, Calif.: RAND, MR-656-JS, 1996.

Porter, Michael, *Competitive Advantage: Creating and Sustaining Superior Performance*, New York: Free Press, 1985.

Schwartz, Peter, *The Art of the Long View*, New York: Doubleday, 1991.

Senge, Peter, *The Fifth Discipline*, New York: Doubleday, 1990.

Smith, Perry, *Creating Strategic Vision: Long-Range Planning for National Security*, Washington, D.C.: National Defense University Press, 1987.

Stacey, Ralph D., *Managing the Unknowable*, San Francisco: Jossey-Bass, 1992.

Stanley, William, "Assessing the Affordability of Fighter Aircraft Modernization," in Davis(1994d), 1994, pp. 565–591.

Thaler, David, *Strategies to Tasks: A Framework for Linking Means to Ends*, Santa Monica, Calif.: RAND, MR-300-AF, 1993.

Thomas, Charles, "Learning from Imagining the Years Ahead," *Planning Review*, May/June 1994.

United States Air Force, *New World Vistas: Air and Space Power for the 21st Century*, Air Force Science Advisory Board, 1996.

Wack, Pierre, "Scenarios: Uncharted Waters Ahead," *Harvard Business Review*, September/October 1985.

_____, "Scenarios: Shooting the Rapids," *Harvard Business Review*, November/December 1985.

Watman, Kenneth, *Asymmetric Strategies for MRCs*, Santa Monica, Calif.: RAND, forthcoming.

Wildavsky, Aaron, "If Planning Is Everything, Perhaps It Is Nothing," *Policy Sciences*, Vol. 4, 1973, pp. 127–153.

Winnefeld, James, *Overseas Presence as a Resource Issue in U.S. Pacific Command: Prototypical Analyses Using RAND's Objectives-to-Programs Methodology*, Santa Monica, Calif.: RAND, unpublished study for the Joint Staff.